Solutions and Tests

for

Exploring Creation

with

General Science
2nd Edition

by
Dr. Jay L. Wile

Solutions and Tests for Exploring Creation With General Science, 2nd Edition

Published by
Apologia Educational Ministries, Inc.
1106 Meridian Plaza, Suite 220
Anderson, IN 46016
www.apologia.com

Manufactured in the United States of America
Second Printing 2008

ISBN: 978-1-932012-87-3

Printed by Courier, Inc., Kendallville, IN

*Cover photos © Stockbyte (boy with test tube), © Sascha Burkard (petrified wood),
© Linda Bucklin (muscle man)*

[Agency for all except Stockbyte: shutterstock.com, Agency for Stockbyte: Getty Images]

Cover design by Kim Williams

Exploring Creation with General Science, 2nd Edition
Solutions and Tests

TABLE OF CONTENTS

Module Tests:

Solutions to the Module Tests:

Quarterly Tests:

Solutions to the Quarterly Tests:

TEACHER'S NOTES
Exploring Creation With General Science, 2nd Edition

Thank you for choosing *Exploring Creation With General Science.* I designed this course to meet the needs of the homeschooling parent. I understand that most homeschooling parents do not know science very well, if at all. As a result, they consider it nearly impossible to teach to their children. This course has several features that make it ideal for such a parent:

1. The course is written in a conversational style. Unlike many authors, I do not get wrapped up in the desire to write formally. As a result, the text is easy to read, and the student feels more like he or she is *learning*, not just reading.

2. The course is completely self-contained. Each module in the student text includes the text of the lesson, experiments to perform, problems to work, and questions to answer. This book contains the solutions to the study guides in the student text, tests, solutions to the tests, and some extra material (answers to the module summaries, cumulative tests, and solutions to the cumulative tests).

3. The experiments are designed for the home. They can be done with items that are readily available at either the grocery store or the hardware store.

4. Most importantly, this course is Christ-centered. In every way possible, I try to make science glorify God. One of the most important things that you and your student should get out of this course is a deeper appreciation for the wonder of God's creation!

Pedagogy of the Text

There are two types of exercises that the student is expected to complete: "On Your Own" problems and study guide questions.

- The "On Your Own" problems should be answered as the student reads the text. The act of working out these problems will cement in the student's mind the concepts he or she is trying to learn. The solutions to these problems are included as a part of the student text. The student should feel free to use those solutions to check his work.

- A study guide is found at the end of each module. It is designed to help the student review what has been covered over the course of the module. It should not be started until *after* the student has completed the module. That way, it will function as a review. It can also be used as a study aid for the test. The student should feel free to use the book while answering the study guide questions.

In addition to the exercises, there is also a test for each module. Those tests are in this book, but a packet of the tests is also included with this book. You can tear the tests out of the packet and give them to your student so that you need not give him this book to administer the tests. You can also purchase additional packets for additional students. You also have our permission to copy the tests out of this book if you would prefer to do that instead of purchasing additional tests for additional students. **I strongly recommend that you administer each test once the student has completed the module**

and the study guide. The student should be allowed to have only a calculator, pencil, and paper while taking the test.

There are also cumulative tests in this book. You can decide whether or not to give these tests to your student. Cumulative tests are probably a good idea if your student is planning to go to college, as he or she will need to get used to taking such tests. There are four cumulative tests along with their solutions. Each cumulative test covers four modules. You have three options as to how you can administer them. You can give each test individually so that the student has four quarterly tests. You can combine the first two quarterly tests and the second two quarterly tests to make two semester tests. You can also combine all four tests to make one end-of-the-year test. If you are giving these tests for the purpose of college preparation, I recommend that you give them as two semester tests, because that is what the student will face in college. The cumulative tests are not in the packet of tests. However, you have our permission to copy them out of this book so you can give them to your student.

Any information that the student must memorize is centered in the text and put in boldface type. Any boldface words (centered or not) are terms with which the student must be familiar. In addition, all definitions presented in the text need to be memorized. Finally, any information required to answer the questions on the study guide must be committed to memory for the test. If the study guide tells the student that he can refer to a particular table or figure in the text, the test will allow him to do so as well. However, if the study guide does not specifically indicate that the student can reference a figure or table, the student will not be able to reference it for the test.

You will notice that every solution contains an underlined section. That is the answer. The rest is simply an explanation of how to get the answer. For questions that require a sentence or paragraph as an answer, the student need not have *exactly* what is in the solution. The basic message of his or her answer, however, has to be the same as the basic message given in the solution.

Experiments

The experiments in this course are designed to be done as the student is reading the text. I recommend that your student keep a notebook of these experiments. The details of how to perform the experiments and how to keep a laboratory notebook are discussed in the "Student Notes" section of the student text. If you go to the course website that is discussed in the "Student Notes" section of the student text, you will also find examples of how the student should record his or her experiments in the laboratory notebook.

Grading

Grading your student is an important part of this course. I recommend that you *correct* the study guide questions, but I do not recommend that you include the student's score in his or her grade. Instead, I recommend that the student's grade be composed solely of test grades and laboratory notebook grades. Here is what I suggest you do:

1. Give the student a grade for each lab that is done. This grade should not reflect the accuracy of the student's results. Rather, it should reflect how well the student followed directions, how well the student recorded his data, and how well he wrote up the lab in his lab notebook.

2. Give the student a grade for each test. In the test solutions, you will see a point value assigned to each problem. If your student answered the problem correctly, he or she should receive the number of points listed. If your student got a portion of the problem correct, he or she should receive a portion of those points. Your student's percentage grade, then, can be calculated as follows:

$$\text{Student's Grade} = \frac{\# \text{ of points received}}{\# \text{ of points possible}} \times 100$$

The number of possible points for each test is listed at the bottom of the solutions.

3. The student's overall grade in the course should be weighted as follows: 35% lab grade and 65% test grade. If you use the cumulative tests, make them worth twice as much as each module test. If you really feel that you must include the study guides in the student's total grade, make the labs worth 35%, the tests worth 55%, and the study guides worth 10%. A straight 90/80/70/60 scale should be used to calculate the student's letter grade. This is typical for most schools. If you have your own grading system, please feel free to use it. This grading system is only a suggestion.

Finally, I must tell you that this course is user-friendly and reasonably understandable. At the same time, however, *it is not EASY*. It is a tough course. Thus, a letter grade of "C" would represent the score of the average student who will most likely be college bound.

Question/Answer Service

For all those who use this curriculum, we offer a question/answer service. If there is anything in the modules that you do not understand – from an esoteric concept to a solution for one of the problems – just contact us via any of the methods listed on the **NEED HELP?** page of the student text. You can also contact us regarding any grading issues that you might have. This is our way of helping you and your student get the maximum benefit from our curriculum.

SOLUTIONS TO THE STUDY GUIDE FOR MODULE #1

1. a. <u>Science</u> – An endeavor dedicated to the accumulation and classification of observable facts in order to formulate general laws about the natural world

b. <u>Papyrus</u> – An ancient form of paper, made from a plant of the same name

c. <u>Spontaneous generation</u> – The idea that living organisms can be spontaneously formed from non-living substances

2. a. <u>We should support a scientific idea based on the evidence, not based on the people who agree with it</u>. Belief in spontaneous generation and the Ptolemaic system lasted so long mostly out of respect for Aristotle and Ptolemy, not because of the evidence.

b. <u>Scientific progress depends not only on scientists, but also on government and culture</u>. Science stalled in the Dark Ages because there was little government or cultural support for it.

c. <u>Scientific progress occurs by building on the work of previous scientists</u>.

3. <u>Imhotep was an ancient Egyptian doctor</u>. His medical practices were renowned throughout the known world.

4. <u>The ancient Egyptians never used their observations to explain the world around them</u>. Instead, they simply took a trial and error approach to finding cures for illness, etc. True science requires observation *and* explanation.

5. <u>They were ancient Greeks who were among the first scientists</u>.

6. They are best remembered for their idea that all matter is composed of <u>atoms</u>.

7. <u>Aristotle</u>

8. <u>Aristotle</u>

9. <u>The geocentric system placed the earth at the center of the universe and had both the planets and the stars traveling around the earth. The heliocentric system has the sun at the center and the planets traveling around it. The heliocentric system is more correct</u>.

10. <u>They wanted to turn lead into gold</u>. "Inexpensive substances" can be substituted for "lead" and "expensive substances" could be substituted for gold.

11. <u>They were not true scientists because their approach was strictly trial and error</u>. They did not try to explain the world around them.

12. <u>Science began to progress towards the end of the Dark Ages because the Christian worldview began to replace the Roman worldview</u>. Since the Christian worldview is a perfect fit with science, the establishment of that worldview was essential for starting scientific progress again.

13. Grosseteste was the first modern scientist because he was first to use the scientific method.

14. The authors were Copernicus and Vesalius. The book by Copernicus was about the arrangement of the stars and planets in space, and the book by Vesalius was on the human body.

15. Galileo was forced to recant belief in the heliocentric system by the Roman Catholic church. Otherwise, he would have been thrown out of the church.

16. Galileo built a telescope, which helped him gather a lot of data about the planets and their motion.

17. Newton was one of the greatest scientists of all time. He laid down the laws of motion, developed a universal law of gravity, invented calculus, wrote many commentaries on the Bible, showed white light is really composed of many different colors of light, and came up with a completely different design for telescopes. The first three are his most important accomplishments in science, but students can list any three.

18. The good part of the change was that science began to stop relying on the authority of past great scientists. The bad part of the change was that science began to move away from the authority of the Bible.

19. Lavoisier came up with the Law of Mass Conservation. He also described combustion, but the Law of Mass Conservation was more important.

20. Dalton is remembered for the first detailed atomic theory.

21. Darwin is best known for his book, *The Origin of Species*. You could also say evolution.

22. The immutability of species refers to the mistaken idea that living creatures cannot change. Darwin showed that this is just not true.

23. Gregor Mendel is remembered for his work on how traits are passed on during reproduction. You could also say genetics.

24. James Clerk Maxwell is known as the founder of modern physics.

25. James Joule came up with The First Law of Thermodynamics.

26. Max Planck first made the assumption that energy comes in small packets called "quanta."

27. Niels Bohr is remembered for his mathematical description of the atom.

28. Einstein also developed the special theory of relativity and the general theory of relativity.

SOLUTIONS TO THE STUDY GUIDE FOR MODULE #2

1. a. <u>Counter example</u> – An example that contradicts a conclusion

b. <u>Hypothesis</u> – An educated guess that attempts to explain an observation or answer a question

c. <u>Theory</u> – A hypothesis that has been tested with a significant amount of data

d. <u>Scientific law</u> – A theory that has been tested by and is consistent with generations of data

2. <u>Science can never prove anything.</u> You could list many of the scientific theories and laws you learned so far that have been demonstrated false in light of new data.

3. <u>No, it does not.</u> You saw that in Experiments 2.1 and 2.2. After all, the idea that heavier things fall faster than lighter things makes sense. Nevertheless, it is wrong!

4. <u>The penny will hit the ground first.</u> Remember, the fact that all things fall at the same rate is only true when there is no air. Air resistance will slow the feather down more than the penny.

5. <u>Neither</u> will hit the bottom of the tube first, because they will both fall at the same rate. Since there is no air in the tube, objects will fall at the same rate, regardless of their weight.

6. To destroy a scientific law, you need <u>only one counter example</u>. Remember, a scientific law is established simply because the theory has been confirmed by an enormous amount of experimentation. If a single experiment can be demonstrated to contradict the law, it is no longer a law!

7. The observation was that <u>the objects similar to the one he was studying had been seen before by other scientists at regular intervals in history</u>.

8. His hypothesis was that <u>the object he was studying was the same thing that the other scientists had seen before</u>.

9. The experiment was <u>to confirm the presence of the comet again in 1758</u>.

10. Since the appearance of the comet has been noted many times throughout history by many different scientists, the existence of Halley's comet is now a <u>scientific law</u>.

11. c. <u>Make observations</u>
a. <u>Form a hypothesis</u>
e. <u>Perform experiments to confirm the hypothesis</u>
d. <u>Hypothesis is now a theory</u>
f. <u>Perform many experiments over several years</u>
b. <u>Theory is now a law</u>

12. <u>You can either discard the hypothesis or modify it to become consistent with the data.</u>

13. <u>You can either discard the theory or modify it to become consistent with the data.</u>

14. The observation that led to Lowell's hypothesis was the fact that there were faint lines on the surface of Mars. The experiments used to confirm the hypothesis were Lowell's detailed studies of Mars' surface, which he thought showed more details about the lines, providing more evidence that his hypothesis was true.

15. The discovery of high-temperature superconductors was startling because a generally-accepted scientific theory (BCS Theory) said it was impossible to have high-temperature superconductors.

16. a. It cannot prove anything.
     b. It is not 100% reliable.
     c. It must conform to the scientific method.

17. Yes, it can. As long as the scientific method is followed, science can be used to study *anything*!

18. Yes, it can. As long as the scientific method is followed, science can be used to study *anything*!

19. The observations were that many people draw strength, hope, and encouragement from the Bible.

20. I hypothesized that the Bible is the Word of God.

21. I searched the Bible for knowledge of future events. This would indicate that the Creator of time had inspired the Book.

22. Of course not! Science cannot prove anything. I did confirm the hypothesis, however. Thus, I provided evidence for its validity.

SOLUTIONS TO THE STUDY GUIDE FOR MODULE #3

1. a. <u>Experimental variable</u> – An aspect of an experiment that changes during the course of the experiment

b. <u>Control (of an experiment)</u> – The variable or part of the experiment to which all others will be compared

c. <u>Blind experiments</u> – Experiments in which the participants do not know whether or not they are a part of the control group

d. <u>Double-blind experiments</u> – Experiments in which neither the participants nor the people analyzing the results know who is in the control group

2. <u>An experimental variable is good when you are using it to learn something from the experiment. An experimental variable should be reduced or eliminated when it affects the results of the experiment but you do not learn anything from it.</u> In Experiment 3.2, for example, the type of "motor" in the "boat" was an experimental variable. It was a good variable, though, because you were using it to learn what kind of motor would work. The other experimental variables should have been reduced or eliminated, because they might have affected the results of the experiment but nothing would be learned from them.

3. The control is <u>the shirt that is being washed with no laundry detergent at all</u>. It is possible that all the detergents are so bad that they have no real effect on the cleanness of the shirts. The only way to tell is to compare it to a shirt that was washed in no detergent.

4. The experimental variable that can be used to learn something from the experiment is <u>the type of detergent used</u>.

5. There are at least four unwanted experimental variables. First, <u>the washers are different</u>. It is possible that some clean clothes better than others. This affects the results of the experiment, because you will not know whether the difference in cleanliness is due to the washer or the detergent. In addition, the <u>water can be at different temperatures</u>, which will affect the outcome. Also, <u>the shirts are different</u>. Some fabrics are easier to clean than others. Finally, <u>the amount of grass stain will be different in each shirt</u>, because there is no way to stain shirts equally.

6. <u>The experimental variable of the washers can be reduced by making sure all washers are the same brand and model, and by making sure they are all relatively new</u>. This will reduce the differences among the washers. <u>You can reduce the differences in water temperature by monitoring the temperature of the water as it enters each washer and making adjustments to keep the temperature the same</u>. <u>The experimental variable of the shirts can be reduced by making sure they are all from the same manufacturer, the same style, and the same fabric</u>. That way, they are as close to identical as possible. Finally, <u>the experimental variable of the amount of stain</u> can be reduced by examining each stain carefully and trying to make sure they are all as identical as possible.

7. <u>The data being collected are subjective</u>. Think about it. Each person's definition of "clean" is different. Also, the shirts are being examined by eye. This makes it hard to say exactly how much stain is left on each shirt. If you could chemically examine each shirt and determine precisely how

much grass stain was left after washing, you would have an objective measurement. However, to have someone just look at a shirt and decide whether or not it is cleaner than another shirt is subjective.

8. The needle floats because of surface tension.

9. Soap reduces the surface tension of water.

10. The liquid must have a larger surface tension than water, because the needle floats more easily on the liquid than it does on water.

11. You should give half the volunteers the fat-free potato chips and the other half should get potato chips that have been on the market for years and seem to have no problems associated with them. The latter group is the control. It's not enough to have the control group eat no potato chips, because the problem might just be with people eating *any* potato chips, not just the fat-free kind. The volunteers then can keep a log (or you could observe them) for the next few hours to see if any stomach cramps occur. If more cramps occur in the group that ate the fat-free chips than what occurred in the control group, the allegations could be true. This should definitely be a double-blind experiment. If the volunteers knew which chips they were getting, it could bias them, and they might imagine stomach cramps when, in fact, they had none. Also, comparing how two groups of people feel after eating is subjective. There is no way to get hard numbers from such a study. Thus, the person analyzing the data needs to be blind as well.

12. This should be a single-blind experiment. If the students knew whether or not they were given the herb, it might influence how they take the test. However, since the data being collected are objective (measurable numbers), there is no reason for the person analyzing the data to not know who is in the control group and who isn't. Of course, if the person wants to avoid any appearance of dishonesty, the experiment could be done as a double-blind experiment.

13. The study should be neither single-blind nor double-blind. The experimental subjects are plants. They cannot "know" whether or not they are a part of the control or not. Thus, the whole idea of a single-blind experiment is kind of irrelevant. Also, the data being collected (the weights of the crops) are objective. The farmer can't bias the result, so there is no need for the farmer to be blind.

14. This should be a double-blind experiment. If the students know whether or not they are in the control group, it might influence how they behave. For example, the homeschooled students might be on their best behavior so as to give the researcher good results for homeschoolers. In the same way, the researcher's observations are subjective. There is no way to precisely measure how one child plays with another. It will depend heavily on the researcher's preconceived notions. Thus, the researcher must be as unbiased as possible and should therefore not know who is in the control group and who is not.

15. This should be a single-blind experiment. The subjects need to be blind because telling a person that he or she is on the "real" drug might affect his or her behavior. For example, a person might eat more than he usually does, assuming that he is "protected" from weight gain. Thus, everyone must be blind as to whether or not they are in the control group. However, there is no need for the researcher to be blind, because the data are completely objective. The people each step on a scale and get weighed. The data, then, are a series of numbers that cannot be affected by the researcher's preconceived notions. Thus, whether or not he knows who is in the control group cannot affect the outcome of the

experiment. Of course, if he is worried about appearing dishonest, he could make it a double-blind study. If he did that, he could not be accused of altering the measurements.

16. If there is no weight on the spring, it will not stretch. Thus, the length of the spring when it is not stretched out will be the length when no weight is hung on it. That means the value of the x-axis is zero. The dot that corresponds to an x-axis value of zero is halfway between the 4 and the 6 on the graph. Thus, the length of the spring with no weight on it is 5 inches. If your answer is not exactly 5 inches, that's okay. Somewhere between 4 inches and 6 inches is fine. Since you are reading from a graph, you do not exactly know where between 4 and 6 the dot is on the y-axis. If you look closely, however, you will see it is halfway between 4 and 6, which means 5.

17. If the spring is stretched to 8 inches, then the dot will be at 8 on the vertical axis. The dot that is on the line for 8 inches is just to the right of 5 pounds, as shown below:

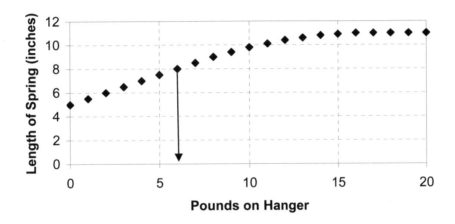

I would say it corresponds to 6 pounds. An answer of 7 would be fine, because you cannot tell exactly. You just know it's between 5 and 10, but much closer to 5 than 10.

18. Notice how the y-values stop increasing after about 15 pounds. After that, no matter how many more pounds are put on the spring, it no longer stretches. Thus, the spring stops stretching after about 15 pounds. Once again, your number could be 14 or 16.

19. If the graphs all have the same shape, it means all the springs stopped stretching after a certain number of pounds were placed on them. Thus, springs stretch in response to a pull, but there is some maximum strength at which they simply no longer stretch.

SOLUTIONS TO THE STUDY GUIDE FOR MODULE #4

1. a. <u>Simple machine</u> – A device that either multiplies or redirects a force

b. <u>Force</u> – A push or pull exerted on an object in an effort to change that object's velocity

c. <u>Mechanical advantage</u> – The amount by which force or motion is magnified in a simple machine

d. <u>Diameter</u> – The length of a straight line that travels from one side of a circle to another and passes through the center of the circle

e. <u>Circumference</u> – The distance around a circle, equal to 3.1416 times the circle's diameter

2. Applied science differs from science in <u>motive</u>. In applied science, the goal is to make something better. In science, the goal is simply to learn.

3. Technology can result from <u>accident, science, or applied science</u>.

4. <u>Experiments (a) and (c) are applied science experiments, because the goal is to make something better. Experiments (b) and (d) are science experiments, because the goal is to learn something.</u> Even though the knowledge gained from experiments (b) and (d) might be useful, that's not the primary goal. Since the primary goal is knowledge, they are science experiments.

5. <u>Items (b) and (c) are technology.</u> They are not machines, but a vaccination for animals is something that makes life better, and a new diet for dogs does the same. Items (a) and (d) are simply pieces of information. They may be useful, but by themselves, they do not make life any better.

6. <u>The lever, the pulley, the wheel and axle, the inclined plane, the wedge, and the screw.</u>

7. <u>The inclined plane and the single wedge</u> look identical.

8. For levers, the mechanical advantage equation (you have to memorize it) is:

Mechanical advantage = (distance from fulcrum to effort) ÷ (distance from fulcrum to resistance)

Mechanical advantage = 40 ÷ 10 = <u>4</u>

9. <u>The mechanical advantage means that the effort is magnified by 4</u>. Remember, in first-class levers, the mechanical advantage magnifies the force.

10. In a shovel, the part that does not move is the handle. You hold onto the handle with one hand, and you lift the middle of the shovel with the other. Thus, the effort is in the middle of the shovel. The resistance is in the shovel's head. Thus, the fulcrum is at one end, and the effort is between the fulcrum and the resistance. This is a <u>third-class lever</u>.

11. In a see-saw, the middle does not move. One child is the resistance and the other is the effort. Thus, the fulcrum is between the effort and the resistance. This is a <u>first-class lever</u>.

12. The mechanical advantage of a wheel and axle is given by:

Mechanical advantage = (diameter of the wheel) ÷ (diameter of the axle)

Mechanical advantage = 15 ÷ 3 = <u>5</u>

13. When the wheel is turned, a wheel and axle magnifies effort. Thus, <u>the applied force will be magnified 5 times</u>.

14. When the axle is turned, speed is magnified. Thus, <u>the wheel will move at 5 times the speed of the axle</u>.

15. The mechanical advantage of a block and tackle is simply equal to the number of pulleys that work together. Thus, the mechanical advantage is <u>6</u>.

16. The mechanical advantage allows you to use less force when you lift, but you "pay" for that by having to pull that much more rope. If the person wants to lift the load 1 foot and the mechanical advantage is 6, the person will need to pull <u>6 feet</u> of rope.

17. The mechanical advantage of an inclined plane is given by:

Mechanical advantage = (length of slope) ÷ (height)

Mechanical advantage = 6 ÷ 2 = <u>3</u>

18. The mechanical advantage equation for a wedge is the same as it is for the inclined plane. Since both have the same dimensions, the mechanical advantage is the same: <u>3</u>.

19. The mechanical advantage equation for a screw depends on the circumference of what is being turned. In this case, it is a screwdriver. Thus, we need to calculate the circumference of the screwdriver:

Circumference = 3.1416 x (diameter)

Circumference = 3.1416 x 2 = 6.2832

Now we can use the mechanical advantage equation for a screw:

Mechanical advantage = (circumference) ÷ pitch

Mechanical advantage = 6.2832 ÷ 0.1 = <u>62.832</u>

20. <u>You should get a fatter screwdriver</u>. The mechanical advantage of a screw/screwdriver combination is dependent on the circumference of the screwdriver. Thus, the fatter the screwdriver, the better!

SOLUTIONS TO THE STUDY GUIDE FOR MODULE #5

1. a. <u>Life science</u> – A term that encompasses all scientific pursuits related to living organisms

b. <u>Archaeology</u> – The study of past human life as revealed by preserved relics

c. <u>Artifacts</u> – Objects made by people, such as tools, weapons, containers, etc.

d. <u>Geology</u> – The study of earth's history as revealed in the rocks that make up the earth

e. <u>Paleontology</u> – The study of life's history as revealed in the preserved remains of once-living organisms

f. <u>Aristotle's dictum</u> – The benefit of the doubt is to be given to the document itself, not assigned by the critic to himself.

g. <u>Known age</u> – The age of an artifact as determined by a date printed on it or a reference to the artifact in a work of history

h. <u>Dendrochronology</u> – The process of counting tree rings to determine the age of a tree

i. <u>Radiometric dating</u> – Using a radioactive process to determine the age of an item

j. <u>Absolute age</u> – The calculated age of an artifact from a specific dating method that is used to determine when the artifact was made

k. <u>The Principle of Superposition</u> – When artifacts are found in rock or earth that is layered, the deeper layers hold the older artifacts.

2. You would use <u>paleontology</u>, because archaeology concentrates on human life.

3. The three tests are: <u>the internal test, the external test, and the bibliographic test</u>.

4. <u>The internal test makes sure that the document does not contradict itself. The external test makes certain that the document does not contradict other known historical or archaeological facts. The bibliographic test makes certain the document we have today is essentially the same as the original</u>.

5. Aristotle's dictum is used in the <u>internal test. We must use it because what seems to be a contradiction in a document might not be a contradiction</u>. It might just be our inability to understand the language in which the document was written.

6. <u>Often those who are making the copy or those who are ordering the copy to be made will make intentional changes</u>. Kings have done this in an effort to make themselves or their ancestors look better in history. Religious groups have been known to do this to make themselves look more important or to make their view look "right."

7. <u>First, there should be a small time period between when the original was written and when the first available copy was made</u>. This reduces the chance for changes being made and reduces the number of

errors that would be committed during the copy process. <u>Second, there must be a lot of different copies from a lot of different sources.</u> If all of the copies agree with one another, we know that a single copier did not make drastic changes.

8. <u>No.</u> The Bible passes the internal test as well as any document of its time.

9. <u>Yes.</u> Because of the difficulty of translating ancient languages, there are some difficult passages. All documents of history have such passages, however.

10. This is a translation problem. The verb "hear" used in Acts 9:7 simply means that the men heard sounds. The verb "hear" used in Acts 22:9 requires that the hearer must actually understand intelligible language. These verses are really complementary, then. <u>The first tells us that the men heard *sounds*, but the second tells us that the men could not *understand* those sounds.</u>

11. <u>One of the genealogies traces Mary's line, while the other traces Joseph's line.</u>

12. We can say this because <u>no other work has had so much archaeological evidence compared to it.</u> The Bible has been tested by archaeology more than any other documents of history, and it passes with flying colors!

13. <u>Sometimes, it turns out that archaeology is wrong</u>, so you cannot discount the validity of a document if archaeology does not fully agree with it. Remember, several archaeologists thought that the Bible was wrong on several occasions. It turns out that it was the archaeologists who were wrong, not the Bible!

14. <u>The New Testament has significantly shorter time spans between original and copy as compared to any other work of the same time period. It also has thousands more supporting documents than any other document of its time.</u>

15. <u>Yes</u>, the Old Testament passes the bibliographic test just as well as any other document of its time.

16. The age is <u>absolute</u>, because a dating method was used to determine it.

17. The coffin has a <u>known age</u>, because it is referenced in a document of history.

18. <u>No.</u> Absolute does not mean certain. Even the most accurate dating method has error in it, and some dating methods can be very unreliable.

19. <u>Master tree ring patterns help the archaeologist determine the age of preserved logs.</u> Master tree ring patterns are cataloged for each region of the world, and they correspond to weather patterns that have already been dated. If an archaeologist finds a master tree ring pattern on a log, he or she knows when that tree ring pattern was formed and can use that to determine the age of the log.

20. The Principle of Superposition assumes that <u>in rock or soil that is layered, the layers were formed one at a time.</u> This is not necessarily true.

21. <u>He can conclude that the city he found was built before 2500 B.C.</u> Assuming the Principle of Superposition is true, the lower layers of rock are older than the upper layers. Since he found this city in a lower layer of rock, it must be older than the city that was discovered in the upper layer of rock.

22. <u>There are many seemingly unrelated cultures that all have a worldwide flood tale</u>. If the flood did not really occur, you have to assume that they all made up the tale independently, because many of the cultures had no contact with one another until well after the tales were written down.

SOLUTIONS TO THE STUDY GUIDE FOR MODULE #6

1. a. <u>Catastrophism</u> – The view that most of earth's geological features are the result of large-scale catastrophes such as floods, volcanic eruptions, etc.

b. <u>Uniformitarianism</u> – The view that most of earth's geological features are the result of slow, gradual processes that have been at work for millions or even billions of years

c. <u>Humus</u> – The decayed remains of once-living creatures

d. <u>Minerals</u> – Inorganic crystalline substances found naturally in the earth

e. <u>Weathering</u> – The process by which rocks are broken down to form sediments

f. <u>Erosion</u> – The process by which rock and soil are broken down and transported away

g. <u>Unconformity</u> – A surface of erosion that separates one layer of rock from another

2. <u>The uniformitarian hypothesis</u> requires that the earth be billions of years old, because it assumes the geological features of the earth took millions and billions of years to form. Catastrophism is more flexible. It can accommodate a young earth or an earth that is billions of years old.

3. The three basic types of rock are <u>igneous, metamorphic, and sedimentary</u>.

4. <u>Igneous rock is the result of molten rock that cools and solidifies. Sedimentary rock is formed when sediments fuse together. Metamorphic rock is formed when either sedimentary or igneous rocks change, usually as a result of temperature or pressure.</u>

5. Most sedimentary rock has been laid down by <u>water</u>.

6. This is <u>physical weathering</u>. The small chips of rock that are broken off are just miniature versions of the original rock. No change in composition has occurred.

7. This is <u>chemical weathering</u>. The limestone forms a gas. That changes the composition of what's left.

8. You expect the most erosion from the <u>quickly-flowing river</u>. Remember, your experiment showed that the faster the water moves, the more erosion that occurs.

9. <u>The barren hillside </u>will experience the most erosion. Remember, your experiment showed that plants reduce the effects of erosion.

10. A river's delta is formed because <u>the river deposits many of the sediments it carries there</u>.

11. Underground caverns are eroded by <u>groundwater</u>.

12. <u>Stalactites form on the ceiling of a cavern, while stalagmites form on the floor of a cavern.</u>

13. Stalactites and stalagmites are formed when groundwater seeps through the ceiling of a cavern. As the drop forms and falls to the floor of the cavern, it might deposit sediments on the ceiling or floor. As those sediments pile up, stalactites and stalagmites are formed.

14. C

15. A and E. Remember, igneous rock is formed from magma. In A, the magma came from a volcanic eruption. In E, it broke into the layers of sedimentary rock.

16. D

17. B

18. E

19. F

20. It is a disconformity.

SOLUTIONS TO THE STUDY GUIDE FOR MODULE #7

1. a. <u>Fossil</u> – The preserved remains of a once-living organism

b. <u>Petrifaction</u> – The conversion of organic material into rock

c. <u>Resin</u> – A thick, slowly flowing liquid produced by plants that can harden into a solid

d. <u>Extinct</u> – A term applied to a species that was once living but now is not

2. The most likely thing that will happen to the remains of a dead plant or animal is that <u>they will decompose</u>. Fossilization is a rare exception to this general rule.

3. <u>A fossil mold forms first</u>. If a cast forms, it forms later when the mold is filled with sediment or magma.

4. <u>The remains of a plant or animal are encased in sediment, and the sediment eventually hardens into rock. As the remains of the plant or animal disintegrate, a hole is left in the rock, in the shape of the original remains. That is the mold. The mold might fill up with sediment or magma later and, when the filling hardens, it forms a cast.</u>

5. Petrifaction requires <u>water that has a lot of minerals in it</u>.

6. Petrified fossils have more information than fossil casts <u>because fossil casts retain only the shape and outer details of the fossil. When a fossil is petrified, its components are replaced with minerals. This means the entire fossil is preserved, which gives us more information than just the shape and outer details of the fossil.</u>

7. <u>No, you cannot.</u> Carbonized remains come as a result of the organism being squished. Thus, they give you a nice two-dimensional view of the organism, but you learn little about its thickness.

8. <u>Plants</u> have tissue that is ideally suited for the carbonization process, so they are the most likely organisms to be fossilized in that way.

9. Fossils encased in amber or ice do not decompose as quickly as other fossils. <u>Thus, tissue and other soft parts tend to be preserved.</u>

10. The four general features of the fossil record are:

I. <u>Fossils are usually found in sedimentary rock. Since most sedimentary rock is laid down by water, it follows that most fossils were laid down by water.</u>
II. <u>The vast majority of the fossil record is made up of hard-shelled creatures like clams. Most of the remaining fossils are of either water-dwelling creatures or insects. Only a tiny, tiny fraction of the fossils we find are of plants, reptiles, birds, and mammals.</u>
III. <u>Many of the fossils we find are of organisms that are still alive today. Many of the fossils we find are of organisms that are now extinct.</u>
IV. <u>The fossils found in one layer of stratified rock can be considerably different from the fossils found in another layer of stratified rock.</u>

11. <u>Clams and other hard-shelled animals</u> make up most of the fossil record.

12. Approximately <u>a thousand</u> species have gone extinct over the last 400 years. This is a stark contrast to the 10,000 species which some "environmentalists" claim go extinct *each year*!

13. A trilobite is <u>a creature that lived in the water and was covered in a hard outer covering.</u> Typically, trilobites lived at the bottom of the ocean. They are now assumed to be <u>extinct</u>.

14. A placoderm is <u>a kind of fish. It was much like the fish we see today, but its head was covered in hard plates rather than scales.</u> Placoderms are considered to be <u>extinct</u>.

15. According to uniformitarians, <u>sediments are laid down slowly over millions of years. Eventually, conditions change and the sediments harden to form rocks. The conditions during which the sediments were laid down determine the type of sediment, which in turn determines the kind of rock formed.</u>

16. According to catastrophists, most of the sedimentary rocks we see today <u>were formed in the worldwide flood. The depth, speed, and direction of the flood waters determined what type of sediments were laid down, which in turn determined the kind of rock formed.</u>

17. According to uniformitarians, each layer of rock represents a period of earth's history. Thus, <u>the different fossils found in different layers result from the fact that different plants and animals existed at different times in any given region.</u>

18. According to catastrophists, most of the sedimentary rock we see today is the result of the worldwide flood. Thus, the depth, speed, and direction of the flood waters determined where the fossils being preserved came from. As a result, <u>the different fossils in different layers are the result of the fact that different kinds of organisms were trapped and preserved during different stages of the flood.</u>

19. Uniformitarians must speculate about <u>how millions of years of time affect the processes we see working today</u>. At best, we have viewed how these processes work over a few thousand years. The effect that millions of years will have on the processes can only be speculated.

20. Catastrophists must speculate about <u>the nature of the worldwide flood</u>. The speculation is aided by the observation of local catastrophes. Nevertheless, the worldwide flood would have been much different from a local catastrophe, so the details of the Flood can only be speculated.

SOLUTIONS TO THE STUDY GUIDE FOR MODULE #8

1. a. <u>Index fossils</u> – Fossils that are assumed to represent a certain period in earth's past

b. <u>Geological column</u> – A theoretical picture in which layers of rock from around the world are meshed together into a single, unbroken record of earth's past

c. <u>The Theory of Evolution</u> – A theory stating that all life on this earth has one (or a few) common ancestor(s) that existed a long time ago

2. Index fossils are used by uniformitarian geologists <u>to determine what time period a layer of rock represents</u>. If a uniformitarian geologist finds index fossils for the Cambrian time period in a layer of rock, for example, the geologist says that the layer of rock was laid down during Cambrian times.

3. The geological column is constructed <u>by comparing layers of rock found in various parts of the world. Using index fossils and the Principle of Superposition, geologists order the layers into one, big column that represents all of earth's geological history</u>.

4. <u>No</u>. The geological column is not something that exists somewhere in the world. It is a construct based on uniformitarian assumptions.

5. Since trilobites are lower on the geological column, uniformitarians assume that <u>trilobites</u> existed on earth before fish.

6. When only algae are found in a layer of rock, the geological column says that the rock is Precambrian, which is assumed to be <u>570 millions years old or older</u>.

7. According to the geological column, fish came before mammals. This means the layer with only fish must be the older layer. Thus, <u>the layer with fish should be on the bottom, and the layer with mammals should be on top</u>.

8. The geological column is viewed as evidence for evolution because <u>it indicates that early in earth's history, there were only simple life forms. As time went on, the geological column indicates that more and more complex life forms started to appear</u>. This is exactly what the Theory of Evolution says.

9. The geological column is not really evidence for evolution because <u>it is not real. Since it is constructed with assumptions, the evidence is only good if the assumptions are valid</u>.

10. The data from Mt. St. Helens indicate that <u>stratified rock can form rapidly</u>.

11. <u>No</u>, you should not make that assumption. The only time we have actually observed a canyon form, it formed as the result of a catastrophe and then later the river formed. Thus, the data from Mt. St. Helens tell us that canyons form rivers, not that rivers form canyons.

12. On a geological scale, the Mt. St. Helens catastrophe was rather <u>minor. This tells us that a major catastrophe would most likely result in larger deposits of stratified rocks and larger canyons</u>.

13. The Mt. St. Helens eruption had <u>mudflows</u>. These mudflows caused most of the geological features that have been studied, and a major flood would also have such mudflows.

14. The Cumberland Bone Cave is a fossil graveyard that contains many fossils from several different climates. This is important <u>because it is excellent evidence for a worldwide flood and is a major problem for the uniformitarian viewpoint.</u>

15. <u>No</u>. Fossilized hats, legs in boots, and waterwheels tell us that fossils can form rapidly.

16. A paraconformity is <u>an unconformity that does not really exist in a geological formation but uniformitarians believe must exist because of the fossils found in the formation.</u>

17. The text discusses the following problems:

a. <u>There are too many fossils in the fossil record.</u>

b. <u>Fossils such as the *Tyrannosaurus rex* bone that contains soft tissue are hard to understand in the uniformitarian framework.</u>

c. <u>Fossil graveyards with fossils from many different climates are hard to understand in the uniformitarian view.</u>

d. <u>Index fossils are called into question by the many creatures we once thought were extinct but we now know are not.</u>

e. <u>Uniformitarians must assume the existence of paraconformities.</u>

f. <u>Uniformitarians must believe that evolution occurred, and there is no evidence for evolution.</u> In fact, the fossil record provides evidence that each plant and animal was created by God.

The student need list only four.

18. The text discusses the following problems:

a. <u>Catastrophists have offered no good explanation for the existence of unconformities between rock layers laid down by the Flood.</u>

b. <u>Catastrophists cannot explain certain fossil structures that look like they were formed under "normal" living conditions which would not exist during the Flood.</u>

c. <u>Catastrophists have not yet explained the enormous chalk deposits we find in terms of the Flood.</u>

19. <u>The fossil record contains no fossils that are undeniable intermediate links.</u> If evolution occurred, there should be many such fossils.

20. <u>The fossil record contains no fossils that are undeniable intermediate links.</u> This is exactly what you would expect if God created each plant and animal individually.

SOLUTIONS TO THE STUDY GUIDE FOR MODULE #9

1. a. <u>Atom</u> – The smallest chemical unit of matter

b. <u>Molecule</u> – Two or more atoms linked together to make a substance with unique properties

c. <u>Photosynthesis</u> – The process by which green plants and some other organisms use the energy of sunlight and simple chemicals to produce their own food

d. <u>Metabolism</u> – The sum total of all processes in an organism that convert energy and matter from outside sources and use that energy and matter to sustain the organism's life functions

e. <u>Receptors</u> – Special structures that allow living organisms to sense the conditions of their internal or external environment

f. <u>Precocial</u> – A term used to describe offspring that are born able to hear, see, move about, regulate body temperature, and eliminate waste without a parent's help

g. <u>Altricial</u> – A term used to describe offspring that are born without at least one of the following abilities: hear, see, move about, regulate body temperature, or eliminate waste

h. <u>Cell</u> – The smallest unit of life in creation

2. The four criteria for life are:

   I. <u>All life forms contain deoxyribonucleic (dee ahk' see rye boh noo klay' ik) acid, which is called DNA.</u>

   II. <u>All life forms have a method by which they extract energy from the surroundings and convert it into energy that sustains them.</u>

   III. <u>All life forms can sense changes in their surroundings and respond to those changes.</u>

   IV. <u>All life forms reproduce.</u>

3. DNA provides the <u>information</u> necessary to turn lifeless chemicals into a living organism.

4. It is <u>big</u>. In fact, DNA is one of the biggest molecules in creation.

5. DNA is <u>significantly more efficient</u> at information storage than the best computer human science can make.

6. a. <u>The nucleotide bases</u> store the information. Remember, the sequence of adenine, thymine, guanine, and cytosine is the code that stores the information.

b. <u>The backbone</u> forms long ribbons that twist to make the double helix structure.

7. <u>Thymine</u> links to adenine, and <u>guanine</u> links to cytosine.

8. The relationship in #7 allows you to determine the other half of the DNA, because only adenine and thymine can link up. Similarly, only cytosine and guanine can link up.

<u>guanine, cytosine, thymine, cytosine, adenine, adenine</u>

9. The relationship in #7 allows you to determine the other half of the DNA, because only adenine and thymine can link up. Similarly, only cytosine and guanine can link up.

<u>adenine, cytosine, guanine, thymine, adenine, cytosine</u>

10. Plants use photosynthesis to make their own food, <u>glucose</u>.

11. Plants often store food as <u>starch</u>.

12. Metabolism requires food and <u>oxygen</u>.

13. Metabolism usually produces energy, <u>carbon dioxide, and water</u>.

14. The organism will not be able to <u>sense and respond to change</u>.

15. <u>The anchovy parents</u> will have many more offspring, because so many anchovies get eaten that many must be born to "replace" them.

16. <u>No, it does not</u>. The cat is still alive, because its cells can reproduce.

17. <u>No, it is not</u>. The speed at which the human population is increasing has been slowing every year. Also, there is more food per person today than ever before. Finally, the cost of raw materials is lower than ever. Thus, there are no indicators that point to trouble.

18. See Figure 9.3

a. <u>organelles</u>
b. <u>nucleus</u>
c. <u>cytoplasm</u>
d. <u>membrane</u>

19. DNA is stored in the <u>nucleus</u>.

20. There are <u>three</u> basic kinds of cells: <u>plant cells, animal cells, and cells from bacteria</u>.

21. The scientist will see <u>two</u> basic kinds of cells. The leaf cell will be a plant cell, and the mouse cell and cat cells will both be animal cells.

SOLUTIONS TO THE STUDY GUIDE FOR MODULE #10

1. a. <u>Prokaryotic cell</u> – A cell that has no distinct, membrane-bounded organelles

b. <u>Eukaryotic cell</u> – A cell with distinct, membrane-bounded organelles

c. <u>Pathogen</u> – An organism that causes disease

d. <u>Decomposers</u> – Organisms that break down the dead remains of other organisms

e. <u>Vegetative reproduction</u> – The process by which one part of a plant can form new roots and develop into a complete plant

2. The five kingdoms are: <u>Monera, Protista, Fungi, Plantae, and Animalia.</u>

3. Since it is made up of several eukaryotic cells, it is not in Kingdom Monera. Since it eats dead organisms, it is probably a decomposer. Thus, it is in kingdom <u>Fungi</u>.

4. Since it is made up of several eukaryotic cells, it is not in kingdom Monera. Since it makes its own food, it is probably a plant. However, you have to be a bit concerned, because it might be an alga and thus belong to kingdom Protista. However, algae do not have specialized structures, so this is not an alga. It therefore must be in kingdom <u>Plantae</u>.

5. If it is a single, prokaryotic cell, it must be in kingdom <u>Monera</u>. The part about eating dead organisms is just there to fool you. Regardless of its eating habits, because it is a prokaryotic cell, it can *only* be a part of kingdom Monera.

6. Since it is made up of several eukaryotic cells and makes its own food, it is either an alga or a plant. Since it has no specialized structures, it is an alga. This puts it in kingdom <u>Protista</u>.

7. If it is a single, eukaryotic cell, it is most likely in kingdom <u>Protista</u>. If it ate only dead organisms, it could be a single-celled fungus. However, the problem says it eats other, living organisms. That means it is not in kingdom Fungi.

8. Since it is made up of several eukaryotic cells, it is not in kingdom Monera. Since it does not make its own food, it is not a plant and thus not in Plantae. Since it is more than one cell and is not an alga, it is not in kingdom Protista. It eats living plants, so it is not in kingdom Fungi, either. Thus, it must be in kingdom <u>Animalia</u>.

9. Bacteria must have water to survive. <u>Dehydrated food has almost all the water removed, so bacteria cannot survive to grow and reproduce.</u>

10. <u>The presence of salt reduces the growth and reproduction of bacteria.</u> Thus, salt protects meat from contamination by bacteria.

11. <u>Bacteria can be introduced onto food or liquid by dust particles in the air. Covering the food keeps the dust particles off, preventing the addition of new bacteria onto the food.</u>

12. If it can move on its own, it is part of the <u>protozoa</u>.

13. <u>Yes</u>, there are many pathogenic organisms in kingdom Protista. The text gives an example.

14. <u>Decomposers recycle the dead matter in creation. Without them, there would be no way that the materials in dead organisms could be used again by living organisms.</u> One reason Biosphere 2 failed was that it didn't have a large enough variety of decomposers.

15. <u>No</u>, not all members of kingdom Fungi are made of several cells. <u>Yeast</u> is an example of single-celled fungus.

16. Like an iceberg, <u>the visible part of a mushroom is actually a small portion of the organism. Most of the organism lies unseen underground.</u>

17. The <u>mycelium</u> is the main part of the mushroom.

18. Only <u>plant cells</u> have cell walls and central vacuoles. It cannot be prokaryotic, because the central vacuole is an organelle.

19. Turgor pressure is the <u>pressure inside of a plant cell that is caused by the central vacuole pushing the cell contents against the cell wall. It allows plants to stand upright.</u>

20. People are members of kingdom <u>Animalia</u>.

SOLUTIONS TO THE STUDY GUIDE FOR MODULE #11

1 a. <u>Axial skeleton</u> – The portion of the skeleton that supports and protects the head, neck, and trunk

b. <u>Appendicular skeleton</u> – The portion of the skeleton that attaches to the axial skeleton and has the limbs attached to it

c. <u>Exoskeleton</u> – A body covering, typically made of chitin, that provides support and protection

d. <u>Symbiosis</u> – A close relationship between two or more species where at least one benefits

2. The human superstructure is made up of <u>the skeleton, the muscles, and the skin</u>.

3. One major difference is the appearance. <u>Under the microscope, smooth muscles appear smooth and unstriped, while skeletal muscles appear rough and striped.</u> The other main difference is the way they operate. <u>Skeletal muscles are voluntary (they are operated by conscious thought), while smooth muscles are involuntary (they are operated unconsciously by the brain).</u>

4. Cardiac muscle is in the <u>heart</u>. It is an <u>involuntary</u> muscle.

5. The red bone marrow produces <u>blood cells</u>.

6. Keratinization is <u>a process that hardens living cells. It is used to make the outer layer of the epidermis, as well as hair and nails.</u> Remember, keratinization kills cells.

7. Bones are made up of <u>collagen and minerals. The collagen makes the bones flexible, while the minerals make them hard and strong.</u>

8. <u>Compact bone tissue is packed together tightly while spongy bone tissue has lots of space in between its fibers.</u> Because of this, spongy bone tissue is lighter than compact bone tissue.

9. <u>Bones are alive.</u> There are living cells imbedded in the bone. Thus, bone tissue is living tissue. That's why it can grow!

10. <u>Vertebrates are animals with backbones. Invertebrates are animals without backbones. It is possible to be neither.</u> If an organism is from *any* kingdom other than Animalia, it is neither an invertebrate nor a vertebrate.

11. According to their definitions, <u>the arms belong to the appendicular skeleton, but the neck is a part of the axial skeleton.</u> See Figure 11.2.

12. <u>An exoskeleton is a support structure that exists on the *outside* of an organism, while an endoskeleton is on the *inside* of the organism.</u> The student could also mention the differences in makeup – endoskeletons are made of bone or cartilage, while exoskeletons are made of chitin. <u>Creatures with exoskeletons are called arthropods.</u>

13. In terms of range of motion, the hinge offers the least, while the ball-and-socket offers the most. Thus, the order in increasing range of motion is: <u>hinge, saddle, ball-and-socket</u>. When there is a larger

range of motion, the joint is less stable. Thus, the order in terms of stability is reversed: ball-and-socket, saddle, hinge.

14. Ligaments hold the bones of the joints together. Cartilage cushions the bones of the joints so that they do not rub painfully against each other.

15. Skeletal muscles end in tendons, and the tendons attach to the skeleton.

16. To tilt the head, the muscles on one side contract, while the muscles on the other side relax. The contracted muscles shorten, pulling on the relaxed muscles. Since the relaxed muscles offer no resistance, they just stretch out.

17. The muscles of the stomach are smooth muscles. You know that because you do not have to think about working your stomach in order for it to do its job.

18. Plants' ability to move towards the light is called phototropism.

19. Hair is used to insulate and provide sensation.

20. Sweat cools the body down and also helps feed the beneficial bacteria and fungi that live on your skin.

21. Skin cells constantly fall off your body because the cells on the outer layer are dead.

22. The sebaceous glands produce oil. This oil softens the skin and hair and also makes it hard for certain bacteria to attach themselves to your skin.

23. Animals with hair are typically mammals. If the skin produces feathers, the animal is a bird. Scales indicate a reptile (or fish, but fish wasn't a choice), and breathing through the skin indicates an amphibian. The last classification is not ironclad. Worms, for example, also breathe through their skin, but they are not amphibians.

    a. mammal
    b. amphibian
    c. reptile
    d. bird

SOLUTIONS TO THE STUDY GUIDE FOR MODULE #12

1. a. <u>Producers</u> – Organisms that produce their own food

b. <u>Consumers</u> – Organisms that eat living producers and/or other consumers for food

c. <u>Herbivore</u> – A consumer that eats producers exclusively

d. <u>Carnivore</u> – A consumer that eats only other consumers

e. <u>Omnivore</u> – A consumer that eats both producers and other consumers

f. <u>Basal metabolic rate</u> – The minimum amount of energy required by the body in a day

2. The energy in most living organisms originates in the <u>sun</u>.

3. a. A mushroom is in kingdom Fungi and is therefore a <u>decomposer</u>.

b. An evergreen bush is a plant and is therefore a <u>producer</u>.

c. A worm eats other things (it is certainly not a plant!), thus it is a <u>consumer</u>.

d. In Module #10, you learned that algae are the most important source of photosynthesis on the planet. Thus, they are <u>producers</u>.

4. Food is converted to energy via the process of <u>combustion</u>.

5. Combustion requires <u>oxygen</u>, as well as something to burn, such as wood or monosaccharides.

6. Combustion produces <u>energy, carbon dioxide, and water</u>.

7. The three macronutrients are <u>carbohydrates, fats, and proteins</u>.

8. The main thing that the macronutrients provide is <u>energy</u>.

9. We need to eat a lot more <u>macronutrients</u>.

10. Disaccharides are made of two monosaccharides linked together, while polysaccharides are made of *several* monosaccharides linked together. Thus, a <u>polysaccharide</u> is the largest.

11. Glucose is a <u>monosaccharide</u>. Most carbohydrates contain a lot of glucose.

12. Fats come in two types: <u>saturated fats</u> and <u>unsaturated fats</u>. You can distinguish them by looking at them while they are at room temperature. <u>Saturated fats are usually solid at room temperature, while unsaturated fats are usually liquid</u>.

13. Proteins are made of long strings of <u>amino acids</u>.

14. The body prefers to burn <u>carbohydrates, then fats, and then proteins (or amino acids)</u>.

15. <u>If your cells do not have enough amino acids, the amino acids from the proteins you eat are sent to your cells so that your cells can make the proteins they need. If your cells have plenty of amino acids, the amino acids from the proteins you eat are either burned for energy or converted into carbohydrates or fats</u>.

16. Your cells must make proteins by linking together amino acids. <u>There are several amino acids your body cannot make. Thus, you must get them from food. Without those amino acids, your cells will not be able to make the proteins they need to make</u>. Animal proteins have these amino acids in plentiful supply. Plant proteins rarely have all of them. Thus, people who eat only plants must get a wide variety of plant proteins to make sure they get those amino acids.

17. <u>Endothermic</u> animals have higher BMRs. The BMR tracks the minimum amount of energy needed to survive. Both endothermic and ectothermic animals have involuntary muscles, etc., that need energy, but only endothermic animals expend energy to keep their internal temperatures high.

18. <u>Ectothermic</u> animals cannot be active on very cold days. Since their body temperatures are not held constant, the colder days reduce the speed at which the chemical reactions can occur in their bodies. This makes them sluggish.

19. Calories are a measure of <u>energy</u>. They can be used to measure how much energy is in food, or how much energy is expended by a living organism.

20. <u>The second man</u> is less active during the day. Since they both burn about the same number of calories while sleeping, they both have essentially the same BMR. Remember, BMR is the minimum amount of energy you need. The amount of energy you burn when you sleep is minimal. If the first needs more calories, he must be burning more energy during the day, when both men are active.

21. <u>No, you cannot</u>. It might be that Jean's BMR is simply much higher than Wanda's. They can each be very active, but their BMR is a major factor in determining how much food they need.

22. In general, the smaller the mammal the higher the normalized BMR, because the more energy the animal has to spend keeping its internal temperature high. Thus, the <u>mouse</u> has the higher normalized BMR.

23. Combustion of food takes place in the <u>cell</u>.

24. The mitochondrion is called the "powerhouse" of the cell because <u>the majority of energy in the combustion process is released in step 3, which takes place in the mitochondrion</u>.

25. The combustion of food takes place in <u>three</u> basic steps. This allows for a <u>gentle release of energy so that the cell doesn't burn up from the combustion process</u>.

SOLUTIONS TO THE STUDY GUIDE FOR MODULE #13

1. a. <u>Digestion</u> – The process by which an organism breaks down its food into small units that can be absorbed by the body

b. <u>Vitamin</u> – A chemical substance the body needs in small amounts to stay healthy

2. a. <u>salivary gland</u> b. <u>epiglottis</u> c. <u>larynx</u> d. <u>trachea</u> e. <u>pancreas</u> f. <u>stomach</u> g. <u>small intestine</u>
h. <u>rectum</u> i. <u>pharynx</u> j. <u>esophagus</u> k. <u>liver</u> l. <u>gall bladder</u> m. <u>large intestine</u> n. <u>anus</u>

3. a. <u>The salivary glands are a part of the digestive system. Technically, they are not part of the digestive tract because food does not pass through them.</u> Don't worry if you got that wrong. <u>They put saliva in the mouth. The saliva partially digests the food, but it also lubricates the mouth and makes it easier for the tongue to form the bolus.</u>

b. <u>The epiglottis is not a part of the digestive system.</u> It is part of the respiratory system.

c. <u>The larynx is not a part of the digestive system.</u>

d. <u>The trachea is not a part of the digestive system.</u>

e. <u>The pancreas is a part of the digestive system but not a part of the digestive tract. It has two main functions. It makes several digestive juices that are squirted into the small intestine as chyme passes through the pyloric sphincter. In addition, it also produces a sodium bicarbonate to neutralize the stomach acid.</u>

f. <u>The stomach is a part of the digestive system and the digestive tract. It churns and mixes the food with gastric juices. The gastric juices contain stomach acid that destroys bacteria that might have been eaten with the food and helps dissolve the food. The gastric juices also contain some digestive chemicals that start the chemical digestion of the food. This turns the bolus of food into chyme.</u>

g. <u>The small intestine is a part of the digestive system and the digestive tract. It chemically digests the food and allows the nutrients to be absorbed through its lining.</u>

h. <u>The rectum is a part of the digestive system and the digestive tract. It pushes feces out of the body through the anus.</u>

i. <u>The pharynx is a part of the digestive system and the digestive tract. It pushes food into the esophagus.</u>

j. <u>The esophagus is a part of the digestive system and the digestive tract. It pushes food down into the stomach.</u>

k. <u>The liver is a part of the digestive system but not a part of the digestive tract. It has many functions. The most important digestion-related functions are making bile, converting glucose into glycogen for storage, breaking down glycogen when the body needs energy, storing fats, and converting parts of the fat and amino acids into glucose when the body needs energy.</u> The liver has many other functions, but those are the ones I want you to remember.

l. The gall bladder is a part of the digestive system but not a part of the digestive tract. It concentrates bile and squirts the bile into the chyme as the chyme enters the small intestine. Bile is a chemical that aids in the digestion of fats and also helps to neutralize the chyme as it enters the small intestine.

m. The large intestine is a part of the digestive system and the digestive tract. It consolidates undigested food, absorbs water from it, and turns the resulting waste into feces. Bacteria in the large intestine also produce vitamin K.

n. The anus is a part of the digestive system and the digestive tract. It is the opening through which feces exit.

4. The appendix acts as a safe haven for beneficial bacteria when sickness can wipe them out of the intestines. This allows the bacteria to repopulate the intestines quickly after the sickness is over.

5. Vitamins A, D, E, and K are fat-soluble.

6. The fat-soluble vitamins are the most likely to build up to toxic levels, as the water-soluble vitamins can be ejected from the body through the urine.

7. Vitamins typically regulate the chemical processes in the cells.

8. Vitamins D and K. Vitamin D is made from sunlight hitting the skin, and vitamin K is made by bacteria in the large intestine.

SOLUTIONS TO THE STUDY GUIDE FOR MODULE #14

1. a. <u>Veins</u> – Blood vessels that carry blood back to the heart

b. <u>Arteries</u> – Blood vessels that carry blood away from the heart

c. <u>Capillaries</u> – Tiny, thin-walled blood vessels that allow the exchange of gases and nutrients between the blood and cells and are located between arteries and veins

2. The lungs <u>oxygenate the blood and allow the blood to get rid of carbon dioxide.</u>

3. The heart <u>pumps blood throughout the circulatory system</u>.

4. The human heart has <u>four chambers: the right atrium, left atrium, right ventricle, and left ventricle.</u>

5. <u>A four-chambered heart is much more efficient than other kinds of hearts.</u>

6. <u>Capillaries</u> are the vessels that allow exchange of gases between the cells and the blood.

7. Deoxygenated blood comes into the heart through the vena cava and gets dumped into the <u>right atrium</u>.

8. After filling the right atrium, the deoxygenated blood is pushed into the <u>right ventricle</u>.

9. When deoxygenated blood leaves the heart, it heads for the <u>lungs</u> to become oxygenated.

10. Oxygenated blood that is returning to the heart fills the <u>left atrium</u>.

11. After entering the left atrium, the oxygenated blood is dumped into the <u>left ventricle</u>.

12. When oxygenated blood is leaving the heart, it is pumped out into the <u>aorta</u>.

13. Veins usually carry <u>deoxygenated blood</u>, because they are carrying it back to the heart to be pumped into the lungs. <u>There are exceptions</u>. The pulmonary vein, for example, carries oxygenated blood to the heart from the lungs. You need not know the name of the exception.

14. <u>The red blood cells carry oxygen to the other cells of the body. The white blood cells fight disease-causing organisms, and the blood platelets aid the blood clotting process.</u>

15. More than half of your blood is blood <u>plasma</u>.

16. <u>Hemoglobin is a protein that carries oxygen. It is found in the red blood cells.</u>

17. Blood cells are produced in <u>bone marrow</u>.

18. <u>Alveoli are small sacs at the end of tiny bronchial tubes. Oxygenation of blood takes place in them. The blood also gives up waste products there.</u>

19. <u>Capillaries surround the alveoli</u>, since that's where the blood gets oxygen and releases wastes.

20. Bronchial tubes are found in the <u>lungs</u>. <u>Air</u> travels through them.

21. <u>nasal cavity, pharynx, larynx, trachea, bronchial tubes, alveoli</u>

22. Vocal cords are found in the <u>larynx</u>.

23. <u>The amount of air that passes over the vocal cords controls the volume of the sound, and the tightness of the vocal cords determines the pitch.</u>

24. The xylem transport water up the plant. The phloem transport food from the leaves to the other parts of the plant. Thus, this sample was taken from the plant's <u>xylem</u>.

SOLUTIONS TO THE STUDY GUIDE FOR MODULE #15

1. a. <u>Gland</u> – A group of cells that prepare and release a chemical for use by the body

b. <u>Vaccine</u> – A weakened or inactive version of a pathogen that stimulates the body's production of antibodies that can destroy the pathogen

c. <u>Hormone</u> – A chemical messenger released into the bloodstream that sends signals to distant cells, causing them to change their behavior in specific ways

2. <u>The lymphatic system fights disease. The urinary system regulates water balance and chemical levels in the blood. The endocrine system controls various functions by releasing hormones.</u>

3. The lymphatic system is made up of <u>lymph vessels and lymph nodes</u>. The lymph gets cleaned in the <u>lymph nodes</u>.

4. <u>The contraction of certain muscles squeezes the lymph vessels, pumping lymph throughout the system.</u>

5. <u>The lacrimal glands produce tears. Tears clean the eye of contaminants and provide a chemical relief for sadness.</u>

6. The lymph nodes have <u>T-cells, B-cells, and macrophages</u> which all fight pathogens and toxic chemicals in different ways.

7. <u>B-cells</u> produce antibodies.

8. <u>Memory B-cells</u> give the lymphatic system a memory of past infections.

9. <u>It does little good</u>. The purpose of a vaccine is to give you the antibodies and the memory B-cells *before* the infection takes place. There are actually a few exceptions to this, because some pathogens "hide" in the body for a while. If you give the vaccine after such an infection, it can stimulate the lymphatic system before the pathogens come out of hiding.

10. <u>Blood enters the kidney via the renal artery. It is filtered to stop the cells and proteins from getting into the nephron. The water and chemicals are dumped into the nephron, and as they travel, cells absorb specific amounts of water and chemicals that then get put back into the blood. Any water and chemicals left over get sent to the renal pelvis and out of the kidney. The cleaned blood leaves through the renal vein.</u>

11. <u>Excess water and chemicals are dumped into the renal pelvis and then travel through the ureter to the bladder. Eventually, they leave the body through the urethra.</u>

12. <u>Dialysis is the process by which a person is hooked up to an artificial kidney when his own kidneys are not functioning properly.</u>

13. <u>d</u>

14. <u>a</u>

15. <u>g</u>

16. <u>e</u>

17. <u>b</u>

18. <u>c</u>

19. <u>f</u>

20. The <u>pituitary gland</u> is often called the "master endocrine gland."

SOLUTIONS TO THE STUDY GUIDE FOR MODULE #16

1. a. <u>Autonomic nervous system</u> – The system of nerves that carries instructions from the CNS to the body's smooth muscles, cardiac muscle, and glands

b. <u>Sensory nervous system</u> – The system of nerves that carries information from the body's receptors to the CNS

c. <u>Somatic motor nervous system</u> – The system of nerves that carries instructions from the CNS to the skeletal muscles

2. <u>Neurons and neuroglia</u> are the two principal kinds of cells in the human nervous system.

3. *See figure 16.3:*

a. <u>dendrites</u>   b. <u>nucleus</u>   c. <u>cell body</u>   d. <u>axon</u>   e. <u>myelin sheath</u>

4. <u>Dendrites carry electrical signals to the cell body</u>.

5. <u>Axons carry electrical signals away from the cell body</u>.

6. <u>A synapse is a small gap between the axon of a neuron and the receiving end of another cell</u>. It is where nerve signals get transferred from one cell to another.

7. <u>When the electrical signal reaches the end of the axon, neurotransmitters are released. They travel across the synapse. Once they reach the receiving cell, they create a new electrical signal</u>.

8. <u>Neuroglia support the neurons by performing tasks that make it possible for the neurons to do their job</u>.

9. <u>The nerve is a part of the PNS</u>. The PNS contains all nerves that run off of the spinal cord. If it is in the leg, it is off the spinal cord.

10. <u>The CNS is made up of the brain and the spinal cord</u>.

11. <u>The skull and the cerebrospinal fluid</u> protect the brain.

12. <u>The vertebral column</u> protects the spinal cord. You could also say "the backbone."

13. Gray matter is made up mostly of <u>neuron cell bodies</u>.

14. White matter is made up mostly of <u>the axons of neurons</u>.

15. The corpus callosum <u>allows the two sides of the brain to communicate with one another</u>.

16. The cerebellum is mostly in charge of <u>skeletal muscle movements</u>, especially fine movements.

17. <u>The cerebrum</u> is in charge of most higher level thinking skills.

18. <u>The two sides of the brain do not do exactly the same things</u>. The left side of the cerebrum, for example, tends to be responsible for speaking, logic, and math. The right side is more involved with spatial relationships, recognition, and music.

19. The left side of the brain controls the <u>right</u> side of the PNS.

20. The blood-brain barrier is <u>a system that "insulates" the brain from the blood. It is important because many of the chemicals in our blood are toxic to brain cells</u>. The blood-brain barrier selectively transports "good" chemicals into the brain and leaves the "bad" chemicals in the capillaries, away from the brain.

21. The sympathetic division <u>increases the rate and strength of the heartbeat and raises the blood pressure. It also stimulates the liver to release more glucose in the blood, producing quick energy for the "fight or flight" response</u> that we experience when we are frightened or angry.

22. The parasympathetic system <u>slows the heart rate and thus lowers the blood pressure. In addition, it takes care of certain "housekeeping" activities such as causing the stomach to churn while it is digesting a meal</u>.

23. Humans detect five basic tastes: <u>salty, sour, sweet, bitter, and umami</u>.

24. When we smell, we are actually detecting <u>chemicals that are in the air</u>.

25. The pupil <u>regulates how much light gets into the eye</u>.

26. The lens <u>focuses light on the retina</u>.

27. The ciliary muscle <u>changes the shape of the lens, adjusting its focus</u>.

28. <u>The rods and cones</u> detect light.

29. <u>Where the optic nerve exits the eye, there are no rods and cones</u>. Without the rods and cones, light cannot be detected on that part of the retina.

30. Experiment 16.5 demonstrated that the <u>fingers</u> are more touch sensitive, which means they have more touch-related receptors.

31. The ear drum <u>converts vibrations in the air into vibrations of the ear ossicles</u>, which send the vibrations on to the cochlea.

32. The cochlea <u>converts the back and forth motion of the ear ossicles into electrical signals that can be received by the brain and interpreted as sound</u>.

<u>Answers to the Module Summaries in Appendix B</u>

Not all of the blanks have to be filled in with exactly the phrases used here. As long as the general message of each paragraph is the same, that's fine.

ANSWERS TO THE SUMMARY OF MODULE #1

1. We should support a scientific idea based on the <u>evidence,</u> not based on the people who agree with it. Scientific progress depends not only on scientists, but also on <u>government</u> and <u>culture</u>. Scientific progress occurs by building on the work of <u>previous scientists</u>.

2. In ancient times, people traveled for miles to visit <u>Imhotep</u> in Egypt, because he was renowned for his knowledge of medicine. Despite the fact that he could cure many ills, his medicine was based not on science, but on <u>trial</u> and <u>error</u>. Egyptian medicine was advanced by the invention of <u>papyrus</u>, which made recording information and passing it on from generation to generation much easier.

3. Three of the first scientists were <u>Thales</u>, <u>Anaximander</u>, and <u>Anaximenes</u>, who were all from ancient Greece. <u>Thales</u> studied the heavens and tried to develop a unifying theme that would explain the movements of the <u>heavenly bodies</u>. His pupil, <u>Anaximander</u> mainly studied life and was probably the first to attempt an explanation for the origin of the human race without reference to a <u>creator</u>. <u>Anaximenes</u> believed that all things were constructed of air, which led to one of the most important scientific ideas introduced by the Greeks: the concept of <u>atoms</u>.

4. <u>Leucippus</u> was a Greek scientist who is known as the father of atomic theory, but the works of his student, <u>Democritus</u>, are much better preserved. This student built on his teacher's foundation, and although most of his ideas about atoms were wrong, he was correct that atoms are in constant <u>motion</u>.

5. <u>Aristotle</u> is often called the father of the life sciences. He was the first to make a large-scale attempt at the <u>classification</u> of animals and plants. Although Aristotle was known for a great number of advances in the sciences, he was also responsible for nonsense that <u>hampered</u> science for many, many years. He believed in <u>spontaneous generation</u>, which says that certain living organisms spontaneously formed from non-living substances. Unfortunately, his <u>reputation</u> caused the idea of spontaneous generation to survive for thousands of years.

6. <u>Ptolemy</u> is best known for proposing the geocentric system of the heavens, where the <u>earth</u> is at the center of the universe and all other heavenly bodies <u>orbit around</u> it. It was later replaced by the more correct <u>heliocentric</u> system, in which the earth and other planets orbit the <u>sun</u>. Three scientists who played a huge role in giving us this system were <u>Copernicus</u>, <u>Kepler</u>, and <u>Galileo</u>. <u>Galileo</u> collected much evidence in support of this system using a <u>telescope</u> he built based on descriptions of a military device. He had to publicly renounce the system, however, for fear of being thrown out of his <u>church</u>.

7. During the Dark ages, <u>alchemy</u> was done in place of science. In this pursuit, people tried to turn lead or other inexpensive items into <u>gold or other valuable metals</u>. These people were not scientists, because they worked strictly by <u>trial</u> and <u>error</u>.

8. Science began to progress towards the end of the Dark Ages because the <u>Christian</u> worldview began to replace the Roman worldview. <u>Grosseteste</u> is generally considered the first modern scientist

because he was first to use the scientific method, although his student, <u>Roger Bacon</u>, is sometimes given that title.

9. In the Renaissance, two very important books were published. One was by <u>Vesalius</u>, and it was a study of the human body. The other was by Copernicus, and it was the first serious proposal of the <u>heliocentric</u> system. In pursuit of data to confirm this system, <u>Kepler</u> was able to develop mathematical equations that showed the planets do not orbit the sun in circles, but in <u>ellipses</u>.

10. <u>Newton</u> was one of the greatest scientists of all time. He laid down the laws of <u>motion</u>, developed a universal law of <u>gravity</u>, invented the mathematical field of <u>calculus</u>, wrote many commentaries on the <u>Bible</u>, showed white light is really composed of many different <u>colors</u> of light, and came up with a completely different design for <u>telescopes</u>.

11. The era of <u>Enlightenment</u> produced good and bad changes for science. The good change was that science began to stop relying on the authority of past <u>scientists</u>. The bad part of the change was that science began to move away from the authority of the <u>Bible</u>. During this era, <u>Linnaeus</u> published his classification system for life, which we still use today. In addition, <u>Lavoisier</u> came up with the Law of Mass Conservation, and <u>Dalton</u> developed the first detailed atomic theory.

12. <u>Darwin</u> is best known for his book, *The Origin of Species*. While most of the ideas in that book have been shown incorrect, it did demonstrate that living organisms can adapt to changes in their surroundings through a process he called <u>natural selection</u>. This essentially destroyed the old, incorrect view called <u>the immutability of species</u>, which says that living creatures cannot change.

13. <u>Louis Pasteur</u> was able to finally destroy the idea of spontaneous generation once and for all. He developed a process called <u>pasteurization</u>, which is used to keep milk from going bad as quickly as it otherwise would. His work laid the foundation for most of today's <u>vaccines</u>, which have saved millions and millions of lives by protecting people from disease.

14. <u>Mendel</u>, an Augustinian monk, devoted much of his life to the study of <u>reproduction</u>. The entire field of modern <u>genetics</u>, which studies how traits are passed on from parent to offspring, is based on his work. Although he loved his scientific pursuits, he gave them up in the latter years of his life because of a political struggle between the government and the <u>church</u>.

15. <u>Maxwell</u> is known as the founder of modern physics, because he was able to show that <u>electricity</u> and <u>magnetism</u> are really just different aspects of the same phenomenon, which is now called <u>electromagnetism</u>.

16. <u>James Joule</u> determined that, like matter, energy cannot be created or destroyed. This is now known as the <u>First Law of Thermodynamics</u>, and it is the guiding principle in the study of energy.

17. In modern era, <u>Planck</u> made the assumption that energy comes in small packets called "quanta." <u>Einstein</u> used that assumption to explain the photoelectric effect, which had puzzled scientists for quite some time. He also developed the special theory of <u>relativity</u> and the general theory of <u>relativity</u>. <u>Bohr</u> also used the assumption that energy comes in quanta to develop a mathematical description for the <u>atom</u>. As a result, the idea that energy comes in little packets is now a central theme in modern science, forming the basis of the theory of <u>quantum</u> mechanics.

ANSWERS TO THE SUMMARY OF MODULE #2

1. Science can never _prove_ anything. However, when the correct _method_ is followed, science can be used to draw _conclusions_ that are reasonably reliable, which can help us better _understand_ the way creation works. A scientific theory does not have to "make sense," it merely has to be consistent with the _data_. A single _counter example_ can destroy a scientific law.

2. In the absence of air, _ALL_ objects, regardless of weight or shape, fall at the _same_ rate. If one object falls slower than another in an experiment, it is most likely the result of air _resistance_. If a penny and a feather are dropped off a cliff, the _penny_ will hit the ground first. If the same experiment is done in an air-free chamber, _neither_ will hit the ground first.

3. The scientific method starts with _observation_, which allows the scientist to collect data. The scientist can then form a _hypothesis_ that attempts to explain some facet of the data or attempts to answer a question that the scientist asks. The scientist then collects much more data in an effort to test the _hypothesis_. If the data are found to be _inconsistent with the hypothesis_, it might be discarded, or it might be modified until it is consistent with the data. If a large amount of data is collected and the _hypothesis_ is consistent with all the data, it becomes a _theory_. If several generations of collected data are all _consistent_ with the _theory_, it eventually attains the status of a scientific _law_.

4. During his career, creation scientist D. Russell Humphreys had read of many observations regarding the earth's magnetic field. They all centered on the fact that ever since scientists have been measuring the earth's magnetic field, it has been weakening. He had also read of measurements of several other planets' magnetic fields. As a result, he developed a mathematical description of how planets form magnetic fields. At this point in the scientific method, his mathematical description would be considered a _hypothesis_. In 1984, he wrote a paper that showed how his mathematical model could accurately reproduce the major observations regarding earth's magnetic field and the measured magnetic fields of the other planets. Even though the magnetic fields of Neptune, Uranus, and Pluto had not yet been measured, he used his mathematical description to predict what they would be if they ever were measured. He also predicted that observations should eventually show that Mars had a magnetic field at one time, but it has long since decayed away. Several years later, the magnetic fields of Neptune and Uranus were measured, and his prediction was the only one that was correct for both planets. In addition, evidence from Mars now indicates that Mars probably did have a magnetic field at one time. As a result, his mathematical description is now a _theory_. He is still using it to make predictions, and if those turn out to be true, eventually, it might become a scientific _law_. Had his 1984 predictions not been validated by the measurements made at Neptune and Uranus, he could have either _discarded_ his mathematical description or _modified_ it until it was consistent with the measurements.

5. Even though he followed the scientific method, _Lowell's_ theory that there were canals on Mars was incorrect. We know now that his experiments were _flawed_ by the lenses used in the telescopes and because of eye strain. Even though BCS theory explained all the available data regarding _superconductivity_, it was shown to be at least partially incorrect when _high-temperature superconductors_ were discovered, because it predicted that no such thing could exist.

6. The failures of the scientific method do not show that science is useless. It shows that science cannot _prove_ anything and is not _100%_ reliable. However, as long as you follow the _scientific method_, you can use it to study anything, and it can produce reasonably _reliable_ conclusions on how the world works.

7. Some say that science cannot be used to study anything we do not observe happening today, but that is <u>wrong</u>. If that were the case, we could not study <u>fossils</u> scientifically, as they are the remains of creatures that lived long ago.

8. Based on historical records, we know the Old Testament existed in its entirety by about <u>250 B.C</u>. Nevertheless, it makes accurate <u>predictions</u> about the fate of cities as well as about the identity and actions of the <u>Messiah</u>.

9. In the book of Ezekiel, there is a prophecy about a city called <u>Tyre</u>. Despite the fact that people during that time thought the city was <u>invulnerable</u>, the prophecy says that God will destroy the city so that it becomes a <u>bare rock</u>, which will only be good for fishermen to use in order to spread their nets. It also says that the city's <u>stones</u> and <u>timbers</u> will be thrown into the water. In fact, both of those predictions came <u>true</u>. Alexander the Great threw the <u>stones</u> and <u>timbers</u> of the mainland city into the ocean to build a bridge of debris to the island city. The once-proud city of Tyre "…is a haven today for fishing boats and a place for <u>spreading nets</u>."

10. The Old Testament properly predicted that Christ would be born in <u>Bethlehem</u> but would later be called out of <u>Egypt</u>. In addition, it properly predicted the <u>amount</u> for which Christ would be betrayed, what Judas would do with the money, and the name of the <u>field</u> the money was used to purchase. According to Josh McDowell, there are at least 332 separate prophecies in the Old Testament that all come true in the life of <u>Christ</u>.

11. Although I gave you <u>evidence</u> that it is scientifically reasonable to believe the Bible, I did *not* <u>prove</u> that the Bible is something in which you should believe. Indeed, since the conclusions of science are always tentative, you should never use it as a basis for your <u>world view</u>.

ANSWERS TO THE SUMMARY OF MODULE #3

1. An experimental variable is good when you are using it to <u>learn</u> something from the experiment. An experimental variable should be reduced or eliminated when it <u>affects</u> the results of the experiment but you do not <u>learn</u> anything from it.

<u>Questions 2-6 are based on the following story</u>:

A consumer lab decides to test the germ-fighting capabilities of different brands of antibacterial soap. The scientists prepare five different dishes, each of which contain the same species of bacteria. In the first dish, no soap is added. In each of the other dishes, a different brand of soap is added and swirled around to mix it with the bacteria in the dish. After 5 minutes, the dishes are examined under a microscope to determine the number of living bacteria in the dish.

2. The control is <u>the dish that has no soap added</u>.

3. The experimental variable from which something will be learned is <u>the soap used in the dish</u>.

4. Lots of experimental variables should be reduced or eliminated. <u>The five different dishes need to be made as identical as possible. The amount of soap added needs to be the same in each case. The amount that each dish is swirled needs to be the same in each case. The person looking at each dish needs to take several breaks and make sure he is examining each dish in exactly the same way so that he counts everything properly. Also, all dishes should be stored in the same place for the 5-minute wait.</u>

5. They are <u>objective</u>, because the bacteria are being counted. That is a solid number that should not depend on the opinion of the person counting.

6. It should be <u>neither</u>. Since the bacteria can't know whether or not they are in the control group, it makes no sense to ask whether or not they should. Also, since the data are objective, there is no need to make the experimenter blind.

7. An object that is denser than water can, under the right circumstances, float on water because of the water's <u>surface tension</u>. However, if soap is added to the water, the object may <u>sink</u>. In general, the stronger the <u>surface tension</u>, the more easily an object of greater density will float on its surface.

8. A student decides to test the effectiveness of an "energy bar" that has been advertised as being able to "unlock" energy reserves, allowing for a more vigorous workout. He gets ten volunteers together. The first five get a bar that looks just like the energy bar, but is really just mashed bread that has been dyed and flavored. This group is the <u>control</u>. The next five get the energy bar. He has all ten do a vigorous workout, and then they each fill out a questionnaire that attempts to find out whether this workout was more vigorous than previous workouts prior to the experiment. This experiment should be done as a <u>double-blind</u> experiment, since the subjects are aware that something is being tested and the data collected are <u>subjective</u>.

9. This should be a <u>double-blind experiment</u>. The people might be affected by knowing whether or not they are getting the "real" treatment. In addition, reading and evaluating people's answers to questions is subjective, so the person evaluating the data needs to be blind.

10. <u>Those who got the fake product</u> are in the control group.

11. <u>The number of days the person fills out a questionnaire</u> is objective, as it is given by simply counting the number of questionnaires. However, since the experimenter is also evaluating the answers to the questions, there is still subjective data, which means the person evaluating the data needs to be blind.

12. This can be a <u>single-blind experiment</u>. The volunteers could be affected by knowing whether or not they are getting the "real" workout. However, the data collection cannot be affected, as the amount of weight a person can lift over his head is not a matter of opinion.

13. During the 16th hour, he caught <u>5 pounds</u> of fish.

14. The <u>7th hour</u> produced the largest number of pounds of fish.

15. <u>Early afternoon</u> is the least productive, as 5 pounds or less of fish are caught each hour. <u>Late evening</u> produces more fish, but <u>early morning</u> gives you the most.

ANSWERS TO THE SUMMARY OF MODULE #4

1. Science is motivated by the desire to <u>understand</u>. In applied science, experiments are aimed at trying to find something <u>useful</u>. Technology is often the result of <u>science</u>, <u>applied science</u>, or <u>accident</u>. Experiments designed to determine why lightning bugs glow, for example, would be considered <u>science</u> experiments. Experiments designed to produce a device that glows like a lighting bug, however, would be <u>applied science</u> experiments. If ever produced, the actual device would be considered <u>technology</u>.

2. A scientist is doing experiments to determine whether or not platinum can speed up certain chemical reactions. Along the way, he discovers two things. He determines that platinum can be used to reduce the toxicity of automobile emissions. This is an example of <u>technology</u>, as it can be used to make people's lives better. He also learns that platinum speeds up chemical reactions by pulling the chemicals close to one another. This is an example of <u>science</u>, because it explains something about how creation works.

3. There are six basic types of simple machines. The <u>lever</u> consists of a bar and a fulcrum. The <u>wheel and axle</u> consists of a wheel that connects to a cylinder. A <u>pulley</u> consists of a wheel over which a rope or chain moves. An <u>inclined plane</u> is simply a ramp. It looks just like a <u>wedge</u>, but the difference between the two is based on where the effort is exerted. The effort is exerted along the <u>slope</u> of an inclined plane, but it is exerted against the <u>short side</u> of a wedge. A <u>screw</u> is composed of an inclined plane wrapped around the cylinder of a <u>wheel and axle</u>.

4. In a first-class lever, the <u>fulcrum</u> is between the effort and resistance. It <u>changes</u> the direction of the effort and magnifies the <u>force</u>. In a second-class lever, the <u>resistance</u> is between the fulcrum and the effort. It <u>does not change</u> the direction of the effort and magnifies the <u>force</u>. In a third-class lever, the <u>effort</u> is between the fulcrum and resistance. It <u>does not change</u> the direction of the effort and magnifies <u>speed</u>. You can determine the mechanical advantage of a lever with the equation:

<u>Mechanical advantage = (distance from fulcrum to effort) ÷ (distance from fulcrum to resistance)</u>

5. A single, fixed pulley offers no <u>mechanical advantage</u>, but it does <u>change</u> the direction of the force. If several pulleys are put together in a block-and-tackle system, there is a <u>mechanical advantage</u>, but you "pay" for it by needing to <u>pull more rope</u>. The force with which you pull on a block and tackle system is multiplied by the <u>number of pulleys</u> in the system.

6. If you turn the wheel of a wheel and axle, the mechanical advantage magnifies the <u>force</u> of the effort. If you turn the axle of a wheel and axle, the mechanical advantage magnifies the <u>speed</u> of the effort. The mechanical advantage of a wheel and axle can be calculated with the equation:

<u>Mechanical advantage = (diameter of the wheel) ÷ (diameter of the axle)</u>

7. In an inclined plane, you can lift an object using less <u>force</u>, but you must push it over a <u>longer</u> distance. A <u>wedge</u> looks just like an inclined plane, but it is used to separate things. You can calculate the mechanical advantage of either with the equation:

<u>Mechanical advantage = (length of the slope) ÷ (height)</u>

8. The mechanical advantage of a screw is calculated with the following equation:

$$\text{Mechanical advantage} = (\text{circumference}) \div (\text{pitch})$$

When grasping the screw's head, you use the circumference of the <u>screw's head</u>. When using a screwdriver, you use the circumference of the <u>screwdriver</u>.

9. The mechanical advantage of a lever is given by:

$$\text{Mechanical advantage} = (\text{distance from fulcrum to effort}) \div (\text{distance from fulcrum to resistance})$$

Plugging the numbers is gives us:

$$\text{Mechanical advantage} = 10 \text{ inches} \div 2 \text{ inches} = \underline{5}$$

10. Since the mechanical advantage of a block and tackle is given by the number of pulleys, this system multiplies force by five. That means you can pull 5 x 50 pounds = <u>250 pounds</u> with it. However, you must pull five times as much rope. Thus, you will need to pull 5 x 10 feet = <u>50 feet</u> of rope.

11. The mechanical advantage of a wheel and axle is:

$$\text{Mechanical advantage} = (\text{diameter of the wheel}) \div (\text{diameter of the axle})$$

$$\text{Mechanical advantage} = (12 \text{ inches}) \div (1 \text{ inch}) = 12$$

This means the force with which you turn the wheel is multiplied by 12. Thus, turning the wheel with 10 pounds of force will produce 12 x 10 pounds = <u>120 pounds</u> of force on the axle. Turning the axle will magnify the speed, so the wheel will turn at 12 x 1 foot per second = <u>12 feet per second</u>.

12. For an inclined plane:

$$\text{Mechanical advantage} = (\text{length of the slope}) \div (\text{height})$$

$$\text{Mechanical advantage} = (10 \text{ feet}) \div (2 \text{ feet}) = \underline{5}$$

13. For a wedge:

$$\text{Mechanical advantage} = (\text{length of the slope}) \div (\text{height})$$

$$\text{Mechanical advantage} = (6 \text{ inches}) \div (1 \text{ inch}) = \underline{6}$$

If the two wedges are put together, the length of the short side will be 2 inches, but the slope will be the same:

$$\text{Mechanical advantage} = (\text{length of the slope}) \div (\text{height})$$

$$\text{Mechanical advantage} = (6 \text{ inches}) \div (2 \text{ inches}) = \underline{3}$$

14. To get the mechanical advantage of a screw, we must know the circumference of what is being grasped. Since no screwdriver is mentioned, it must be the head:

$$Circumference = 3.1416 \times (diameter)$$

$$Circumference = 3.1416 \times 0.5 = 1.5708$$

Now we can use the mechanical advantage equation for a screw:

$$Mechanical\ advantage = (circumference) \div pitch$$

$$Mechanical\ advantage = 1.5708 \div 0.1 = \underline{15.708}$$

15. Now the circumference used in the mechanical advantage equation comes from the screwdriver:

$$Circumference = 3.1416 \times (diameter)$$

$$Circumference = 3.1416 \times 1 = 3.1416$$

Now we can use the mechanical advantage equation for a screw:

$$Mechanical\ advantage = (circumference) \div pitch$$

$$Mechanical\ advantage = 3.1416 \div 0.1 = \underline{31.416}$$

If you used a 2-inch diameter screwdriver for the screw shown above, what would the mechanical advantage be?

ANSWERS TO THE SUMMARY OF MODULE #5

1. When the history of life is studied, <u>paleontology</u> examines all life that once existed on the planet, while <u>archaeology</u> concentrates on human life. Studying rocks to learn the history of the earth is called <u>geology</u>.

2. The three main tests used to determine whether or not a document is a valid work of history are called <u>the internal test</u>, <u>the external test</u>, and <u>the bibliographic test</u>. The <u>internal test</u> makes sure that the document does not contradict itself. The <u>external test</u> makes certain that the document does not contradict other known historical or archaeological facts. The <u>bibliographic test</u> makes certain the documents we have today are essentially the same as the original.

3. <u>Aristotle's dictum</u> is used in the internal test. We must use it because what seems to be a contradiction in a document might not be a contradiction. It might just be the result of the fact that we cannot fully <u>understand</u> the language in which the document was written.

4. In the bibliographic test, a document is considered reliable if there is a <u>short</u> time between when the original was written and the oldest copy that exists today. In addition, the <u>larger</u> the number of copies made by <u>different</u> sources, the more reliable the document. This is important, because the people who could commission copies of a work might also demand that <u>changes</u> be made to the original. In addition, because copying was a tedious process, copyists sometimes made <u>mistakes</u> when they made copies.

5. The Bible <u>passes</u> the internal test as well as any document of its time. Most "contradictions" that people claim exist in the Bible are not contradictions at all. For example, while the genealogies of Christ in Matthew and Luke seem contradictory, they are actually complimentary, because Luke traces <u>Mary's</u> line, while Matthew traces <u>Joseph's</u> line. In addition, while the stories of Paul's conversion in Acts 9 and 22 appear to be contradictory, they do not contradict each other in the original language. The verb "hear" used in Acts 9:7 simply means that the men heard <u>sounds</u>. The verb "hear" used in Acts 22:9 requires that the hearer must actually understand <u>language</u>. The first tells us that the men heard <u>sounds</u>, but the second tells us that the men could not <u>understand</u> those sounds as speech.

6. While there are a few difficult passages in the Bible, the same can be said of any <u>ancient</u> historical document. Thus, unless you can <u>conclusively</u> demonstrate a contradiction, it is labeled a difficulty and does not cause a document to fail the <u>internal test</u>.

7. The Bible <u>passes</u> the external test. In fact, since more archaeology has been done in relation to the Bible than any other work of history, you can say it passes the external test <u>better</u> than any other document of ancient history. In fact, the few times in which archaeology was thought to contradict the Bible, it turned out that <u>archaeology</u> was wrong, not the Bible.

8. The New Testament <u>passes</u> the bibliographic test <u>better</u> than any other ancient document of its time. The time span between when the original was written and the first copy that exists today is very <u>short</u>, and the number of independent copies of the manuscript is very <u>large</u>.

9. The Old Testament <u>passes</u> the bibliographic test. One of the most important finds related to this was the discovery of the <u>Dead Sea Scrolls</u>. While there were several works in this group of documents, there was a complete copy of the book of <u>Isaiah</u>.

10. In archaeology, a document or relic can have a <u>known</u> age, an <u>absolute</u> age, or a <u>relative</u> age. An object has a <u>known</u> age if the date appears on the object, or if the object is referenced in some other valid work of history. It has an <u>absolute</u> age if a dating method is used to calculate its age. It has a <u>relative</u> age if the Principle of Superposition is used to determine whether it is older or younger than some other object. Only <u>known</u> ages are reasonably certain. The others rely on assumptions.

11. The Principle of Superposition says that the <u>lower</u> you find a relic in a structure of layered rock or soil, the older the object is. Thus, if an archaeologist finds pottery in one layer of soil and bones in a layer below, he can say the bones are <u>older</u> than the pottery. This principle assumes that rock and soil layers form <u>one at a time</u>. While this is certainly true for certain situations, it has been demonstrated to be <u>false</u> for other situations, so the Principle of Superposition is not very reliable.

12. The counting of tree rings in order to determine the age of something is called <u>dendrochronology</u>. Typically, such ages are considered <u>upper</u> limits on the true age, as some trees can grow more than one ring in a given year. In order to determine the absolute age of a log that was cut down and then preserved, the investigator looks for a <u>master tree ring pattern</u>. This pattern corresponds to a series of weather conditions for which an absolute age has already been determined.

13. In addition to dendrochronology, radiometric dating can be used to determine the <u>absolute</u> age of an artifact. While most forms of this dating technique are unreliable, the radiometric technique known as <u>carbon-14 dating</u> is about as reliable as dendrochronology as long as the item being dated is less than 3,000 years old.

14. There are many seemingly unrelated cultures that all have a <u>worldwide flood</u> tale. If the Flood did not really occur, you have to assume that they all <u>made up</u> the tale independently, because many of the cultures had no contact with one another until well after the tales were written down. One of the more famous examples is the Epic of <u>Gilgamesh</u>. This story details the adventures of a king who eventually seeks a wise man who <u>survived</u> the worldwide flood. The man tells the king about the Flood and how to become young again. The king <u>fails</u> the test needed to eat the plant that makes him young, however.

15. While there are many ways to study human history, the best place to start is the <u>Bible</u>, because it has been shown to be an incredibly reliable source of history.

ANSWERS TO THE SUMMARY OF MODULE #6

1. There are two basic viewpoints when it comes to forming hypotheses about earth's past: catastrophism (which assumes the majority of the geological record is the result of catastrophes that have happened during earth's past) and uniformitarianism (which assumes the majority of the geological record is the result of processes that we see happening continuously today). In general, uniformitarians are forced to believe that the earth is billions of years old, while catastrophists can be more open-minded.

2. There are three basic kinds of rocks that make up the earth's crust: sedimentary rock (formed from sediments), igneous rock (formed from magma), and metamorphic rock (formed from either of the first two kinds of rock). Because of the changes that form metamorphic rock, it is typically very hard.

3. The basic building blocks of rocks are called minerals, and they can often be distinguished from rocks by their nice crystalline shapes.

4. Sedimentary rock often forms in layers, which are called strata. Most sedimentary strata are laid down by water. These layers vary greatly in thickness and appearance.

5. When molten rock is found under the surface of the earth, it is usually called magma. When it is found on the surface (because of a volcanic eruption, for example), it is called lava.

6. Rocks can experience physical weathering, where they are broken into small pieces, or chemical weathering, where they are transformed into new substances. For example, if a rock is worn away from the constant pummeling of raindrops, it has experience physical weathering. On the other hand, when a rock falls apart because the iron inside has been changed to rust, it has experienced chemical weathering. Either way, weathering transforms rock into sediments. Generally, the process of weathering takes a long time. Thus, it is hard to see it happening on a day-by-day basis.

7. Erosion is responsible for shaping geological structures and landscapes into what we see today. While running water (like that found in a river) is the most common agent of erosion, wind and rain can also be its agents. In general, the presence of plants tends to slow erosion, because their roots hold the soil together. In addition, the swiftly-flowing water erodes things faster. In general, erosion is one way creation recycles its rocks, turning rocks into sediment and then depositing the sediments somewhere, forming rocks again.

8. When a river erodes a landscape, the sediments are usually carried along and deposited wherever the river ends. As a result, a river's end is often a fan-shaped area of deposited sediments called a delta. While we normally think of rivers as the agents of erosion, groundwater (water flowing beneath the earth's surface) can also erode rock. We can see the effects of this erosion by looking at a cavern.

9. In a cavern, water dripping from the ceiling can deposit sediments, forming icicle-like structures called stalactites. When water hits the ground, it can deposit sediments there, causing a stalagmite to rise up from the ground. If an icicle-like structure meets a structure rising from the ground, the result is a column, which is also called a pillar. While caves are the most common place to find these structures, they can form in man-made basements as well.

10. When strata are separated by a surface that has been eroded, we call that separation an <u>unconformity</u>. If the separation is between sedimentary rock and either igneous or metamorphic rock, it is called a <u>nonconformity</u>. If the separation is between two parallel layers of sedimentary rock, it is called a <u>disconformity</u>. If the layers above and below the separation are tilted relative to one another, it is called an <u>angular unconformity</u>. If geologists think there should be an unconformity but there is no evidence for one, it is called a <u>paraconformity</u>. Not all layers of rock are separated by <u>unconformities</u>. If a layer of rock rests on another layer with no evidence of erosion in between, the separation is not an <u>unconformity</u>.

12. Veins of igneous rock that shoot through layers of sedimentary rock are called <u>intrusions</u>. Veins that run in the same direction of the strata are called <u>sills</u>, while veins running roughly perpendicular to the direction of the strata are called <u>dikes</u>.

13. In the side-on view of the Grand Canyon given in the book, letters A, E, and F point to <u>igneous</u> rock. While the rock pointed out by A formed from <u>lava</u>, the rock pointed out by E and F formed from <u>magma</u>. Letters E and F point to an <u>intrusion</u>, and specifically, F points to a <u>sill</u> while E points to a <u>dike</u>. The unconformity pointed out by D is called the <u>Great Unconformity</u>, and the type of unconformity is an <u>angular unconformity</u>. The layers above D are all composed of <u>sedimentary</u> rock. Letter H points to <u>metamorphic</u> rock, which makes the unconformity pointed to by the letter G a <u>nonconformity</u>. The unconformity pointed to by letter C is a <u>disconformity</u>. If a geologist is studying the layer pointed out by B and thinks there should be an unconformity roughly in the middle of the layer despite the fact that there is no evidence for one, it would be called a <u>paraconformity</u>.

ANSWERS TO THE SUMMARY OF MODULE #7

1. In general, dead organisms <u>decompose</u>. Only a tiny fraction of once-living organisms gets fossilized. As a result, the conditions under which fossilization form are <u>rare</u>.

2. The most common means by which a dead organism can be preserved is by the formation of a <u>mold</u> and the making of a <u>cast</u>. In this process, the <u>organism itself</u> is not actually preserved. Instead, rocks are formed in the <u>outer details</u> of the organism.

3. Sometimes, a mold will form without a cast. When that happens, an <u>impression</u> of the organism is left in the rock, but nothing fills it.

4. <u>Petrifaction</u> is the process by which the organic materials in an organism are replaced by minerals. This process requires <u>mineral-rich</u> water. These fossils have more information than fossil casts and molds because the <u>entire</u> fossil is preserved, which gives us more information than just the shape and outer details of the fossil.

5. When an organism is buried in sediment, the pressure can cause <u>liquids</u> and <u>gases</u> in the organism's remains to be forced out into the surrounding sediments. This means the <u>majority</u> of the organism's remains are lost. A process called <u>carbonization</u> takes place, leaving a thin, filmy residue that often forms a detailed "drawing" of the creature in stone. <u>Plant</u> fossils are the most common fossils created by this process. While this process can create very detailed "drawings" of the organism, it does not preserve the details of the organism's <u>thickness</u>, as the "drawings" are flat.

6. The fossils that contain the most information are those that <u>avoided</u> decomposition. This can happen, for example, which an organism is encased in <u>ice</u>. The cold temperatures and protection offered by the <u>ice</u> slow decomposition so much that organic remains can be preserved. If an organism is encased in <u>amber</u>, a hardened resin, it can also be well-preserved. However, even though such a fossil appears to be intact, the <u>chemicals</u> that make it up have been decomposing since the organism died, so the organism is not fully preserved.

7. There are four general features of the fossil record. They are as follows:

    a. <u>Fossils are usually found in sedimentary rock. Since most sedimentary rock is laid down by water, it follows that most fossils were laid down by water.</u>

    b. <u>The vast majority of the fossil record is made up of clams and other hard-shelled creatures. Most of the remaining fossils are of either water-dwelling creatures or insects. Only a tiny, tiny fraction of the fossils we find are of plants, reptiles, birds, and mammals.</u>

    c. <u>Many of the fossils we find are of organisms that are still alive today. Many of the fossils we find are of organisms that are now extinct.</u>

    d. <u>The fossils found in one layer of stratified rock can be considerably different from the fossils found in another layer of stratified rock.</u>

8. When paleontologists compare a fossil to its living counterpart (assuming one exists), they find that the fossil and the living counterpart are incredibly <u>similar</u>. Based on the fossil evidence, then, we can conclude that organisms that have survived throughout earth's history experience little <u>change</u>. This does not mean organisms experience no <u>change</u> at all. However, the <u>changes</u> are minor compared to the characteristics that define the organism.

9. When an organism is found in the fossil record, but no living counterpart is found, it is assumed that the organism is <u>extinct</u>. That assumption, however, is sometimes <u>wrong</u>. From time to time, biologists will find a living example of an organism thought to be <u>extinct</u>.

10. There is a lot of misinformation regarding extinction. While some claim that tens of thousands of organisms go extinct every year, scientists estimate that about <u>a thousand</u> have gone extinct in the past 400 years.

11. Several fossils can be found in the Grand Canyon. There are fossils of <u>trilobites</u> (mostly bottom-dwelling ocean creatures) in many of the layers above the Great Unconformity. They are assumed to be <u>extinct</u>. In the three layers just above the Great Unconformity, there are no direct fossils of worms, but we find fossils of their <u>burrows</u>, indicating their presence. In only one layer of the Grand Canyon, you will find fossils of <u>placoderms</u> (fish that had bony plates covering their heads rather than the scales you see covering the heads of fish today). They are also assumed to be <u>extinct</u>. Fish that look like the ones we see today are found only in the <u>topmost</u> layers of the Grand Canyon.

12. According to uniformitarians, sediments are laid down <u>slowly</u> over <u>millions</u> of years. Eventually, conditions change and the sediments harden to form <u>rock</u>. The conditions during which the sediments were laid down determine the <u>type</u> of sediment, which in turn determines the <u>kind</u> of rock formed. A phrase that best sums up the uniformitarian position is, "The <u>present</u> is the key to the past." According to uniformitarians, each layer of rock represents a <u>period</u> of earth's history. Thus, the different fossils found in different layers result from the fact that different plants and animals existed at <u>different times</u> in any given region of the earth.

13. According to catastrophists, most of the sedimentary rocks we see today were formed during the <u>worldwide flood</u>. The depth, speed, and direction of the Flood waters determined what <u>type</u> of sediment was laid down, which in turn determined the <u>kind</u> of rock formed. As a result, the different fossils in different layers are the result of the fact that different kinds of organisms were trapped and preserved during different <u>stages</u> of the Flood.

14. Both uniformitarianism and catastrophism require speculation. Uniformitarians must speculate how <u>millions of years</u> of time affect the processes we see working today. Catastrophists must speculate about the nature of the <u>worldwide flood</u>.

ANSWERS TO THE SUMMARY OF MODULE #8

1. Uniformitarian geologists use <u>index fossils</u> to determine what time period a layer of rock represents. If a uniformitarian geologist finds <u>index fossils</u> for the Cambrian time period in a layer of rock, for example, the geologist says that the layer of rock was laid down during Cambrian times.

2. Uniformitarian geologists assume that not every time period of earth's history will be represented by rock in every part of the world. In a given part of the world at a give time, the conditions for <u>sediment</u> deposition might not have existed. Also, rock might have been laid down in a certain region, but it might have <u>eroded</u> before another layer of rock could form. They use geological data from all over the world to form the <u>geological column</u>, a theoretical picture of earth's entire history. It assumes that each layer of rock represents a <u>period</u> in earth's past, and it further assumes that the index fossils found in a given layer of rock are accurate indicators of which time <u>period</u> the rock was formed. If either of these assumptions is wrong, the geological column is probably not <u>accurate</u>.

3. In the geological column, you find trilobites and similar animals much lower than you find mammals. According to uniformitarian geologists, this is because trilobites existed <u>before</u> mammals. In the same way, you generally find the fossils of mammals higher in the geological column than dinosaurs. According to uniformitarian geologists, this is because dinosaurs existed <u>before</u> mammals.

4. Many view the geological column as evidence for <u>evolution</u>, because it indicates that early in earth's history, there were only simple life forms. As time went on, the geological column indicates that more and more <u>complex</u> life forms started to appear. This is exactly what the Theory of Evolution says. The problem with using the geological column as evidence for evolution is that the geological column is not <u>real</u>. It is an abstract model based on <u>assumptions</u> that may or may not be correct.

5. We already know that the geological column is <u>wrong</u> to some extent. This is because over time, paleontologists have found fossils in <u>Cambrian</u> rock that, according to the geological column, should not have existed until much <u>later</u> in earth's history. A more accurate geological column would have trilobites and similar animals, as well as ocean life without bones, all together in Cambrian, Ordovician, and Silurian rock. Unfortunately, most textbooks <u>do not</u> present such an accurate picture.

6. The Theory of Evolution states that as millions of years pass and life forms reproduce, small <u>differences</u> between parent and offspring appear. These small differences can "pile up" over time until there are so many differences that the offspring being produced look nothing like the life form that <u>began</u> this process. In this way, a "<u>simple</u>" life form can give rise to a more <u>complicated</u> life form. This happened over and over again as time went on, producing more and more <u>complex</u> organisms over earth's history.

7. During the eruption of Mount Saint Helens, there were periods when huge volumes of <u>steam</u> were released. This ground-hugging <u>steam</u>, mixed with volcanic ash, formed a "river" of mud that moved across the ground at speeds greater than 100 miles per hour. As <u>sediments</u> were deposited by this "river," they formed <u>layers</u>, which were laid down with varying thickness. All the <u>layers</u> were deposited together, in the span of just a few hours!

8. The eruption of Mt. Saint Helens also showed how quickly fast-moving water can <u>erode</u> rock. A huge canyon was carved out of solid rock as a result of the <u>mudflows</u> that accompanied the eruption. The formation of the canyon also caused the formation of a <u>river</u> at the bottom of the canyon.

9. If the Mt. Saint Helens catastrophe can build layers of stratified sediments that are several feet high, it stands to reason that a <u>worldwide flood</u> could form layers of stratified sediments that are hundreds or thousands of feet high. If the Mt. Saint Helens catastrophe can carve out canyons that are <u>one-fortieth</u> the scale of the Grand Canyon, a large, post-flood catastrophe could certainly carve out the Grand Canyon itself.

10. The Cumberland Bone Cave is a fossil <u>graveyard</u> that contains many fossils from several different climates. It is excellent evidence for a <u>worldwide flood</u> and is a problem for the uniformitarian view.

11. While it is commonly assumed that fossils take thousands, if not millions, of <u>years</u> to form, it is not necessarily true. Fossilized hats, legs in boots, and waterwheels tell us that fossils can form <u>rapidly</u> under the right conditions. In addition, there are museums that carry carbonized remains of a large fish in the process of <u>eating</u> a smaller fish. The best way to understand such a fossil is to realize that the fish were buried in an <u>instant,</u> without warning. This killed both the fish and its potential meal. Since they were buried by sediment, the process of forming a <u>carbonized</u> fossil began.

12. A <u>paraconformity</u> is an unconformity that does not really exist in a geological formation but uniformitarians believe must exist because of the fossils found in the formation. This is one of several problems with the uniformitarian view. Another problem is that there are <u>too many</u> fossils in the fossil record. In addition, fossil <u>graveyards</u> with fossils from many different climates are hard to understand in the uniformitarian view. Another problem comes from fossils such as the *Tyrannosaurus rex* bone containing <u>soft</u> tissue that should not have lasted for millions of years. Also, the entire idea of using <u>index fossils</u> to order sedimentary strata is called into question by the many creatures we once thought were extinct but we now know are not.

13. Catastrophists have offered no good explanation for <u>unconformities</u> between rock layers laid down by the Flood. Another problem with the catastrophism framework is the existence of fossil structures that look like they were formed under <u>normal</u> living conditions, which would not exist during the Flood. In addition, catastrophists have not yet explained the enormous <u>chalk</u> deposits we find in terms of the Flood.

14. The fossil record contains no fossils that are undeniable <u>intermediate</u> links. If evolution occurred, there should be <u>many</u> such fossils. If God created each kind of plant and animal individually, you would <u>not</u> expect any.

15. Evolutionists think that *Archaeopteryx* is an intermediate link between reptiles and birds. They think this because it has <u>teeth</u> in its mouth. No living bird has them, but reptiles do. In addition, there are <u>claws</u> on the wings. No living adult bird has them, but reptiles do. The <u>tail</u> is also longer and has more bones than that of any living bird we see today. Detailed studies of the fossil, however, indicate that the animal was an excellent <u>flyer</u>, as you would expect of a fully-developed bird. Thus it seems to be just a bird with certain <u>special features</u> that no living bird has today.

16. There are <u>creationists</u> who are also uniformitarians. They believe in the uniformitarian view of geology, but they agree that the Theory of Evolution is not a valid explanation of life's history on earth. One such group is referred to as progressive <u>creationists</u>. In this view, God created <u>simple</u> creatures, allowed them to live, reproduce, etc., and then, after a long while, He created slightly more <u>complex</u> creatures. Once again, He then paused and allowed them to "do their thing" for a long while, and then He created even more <u>complex</u> creatures. After enough time, of course, this would produce all the basic kinds of organisms we see today.

ANSWERS TO THE SUMMARY OF MODULE #9

1. All life forms contain deoxyribonucleic acid, which is called <u>DNA</u>. All life forms have a method by which they extract <u>energy</u> from their surroundings and convert it into <u>energy</u> that sustains them. All life forms can sense <u>changes</u> in their surroundings and <u>respond</u> to them. All life forms <u>reproduce</u>.

2. DNA provides the <u>information</u> necessary to turn lifeless chemicals into a living organism. It is one of the <u>biggest</u> molecules in creation, and it is more efficient at <u>information</u> storage than the very best that human technology has to offer.

3. A DNA molecule is shaped like a double <u>helix</u>. It is formed by two long strands of atoms that make up the <u>backbone</u> of the DNA. Each strand has little units, called <u>nucleotide bases</u>, attached to it. The units come in four varieties: <u>adenine</u>, <u>thymine</u>, <u>guanine</u>, and <u>cytosine</u>. The information is stored in the <u>sequence</u> of these little units.

4. The double helix of DNA stays together because the nucleotide bases can link together. However, only adenine and <u>thymine</u> can link together, and only cytosine and <u>guanine</u> can link together. As a result, if one strand of DNA has the following sequence of nucleotide bases:

adenine, cytosine, cytosine, thymine, guanine, guanine, thymine

The other strand must have the following sequence of nucleotide bases:

<u>thymine</u>, <u>guanine</u>, <u>guanine</u>, <u>adenine</u>, <u>cytosine</u>, <u>cytosine</u>, <u>adenine</u>

5. Plants use <u>photosynthesis</u> to produce food in the form of a chemical called <u>glucose</u>. If the plant needs energy, it takes the energy from the <u>glucose</u> and converts it into energy it uses to survive. If the plant has plenty of energy, it converts any unused <u>glucose</u> into <u>starch</u> (or one of a few other chemicals), which can be broken back down into <u>glucose</u> when the plant needs energy.

6. The process of photosynthesis uses <u>carbon dioxide</u>, <u>water</u>, and energy from the <u>sun</u> to produce <u>glucose</u> and oxygen. Even though plants produce oxygen, they also <u>use</u> oxygen. When they want to get the <u>energy</u> from the glucose, they must perform <u>metabolism</u>, which requires food (like glucose) and oxygen. However, since plants <u>make</u> more food than they ever <u>use</u>, they <u>make</u> more oxygen than they <u>use</u>.

7. Living organisms are equipped with some method of receiving information about their <u>surroundings</u>. Typically, they accomplish this feat with <u>receptors</u>. This allows them to <u>adapt</u> to any changes that occur. People with leprosy cannot feel <u>pain</u>, and as a result, they tend to have real troubles when they are hurt. Their wounds get <u>infected</u>, and they can lose body parts because they do not know they have been hurt. The disease known as leprosy today, however, is not the disease the <u>Bible</u> calls leprosy. Most likely the disease called leprosy in the <u>Bible</u> was a variety of skin disorders that were probably caused by an infectious agent.

8. Since plants are living organisms, they can <u>respond</u> to changes in their environment. One classic example of this is the fact that plants can grow <u>toward</u> a light source to get as much energy as possible from the light.

9. Reproduction is a means by which living organisms ensure that their kind will continue. Of course, sometimes this is not enough. As we learned from geology and paleontology, some organisms go extinct. This can be for a variety of reasons, but in the end, it comes down to one thing: the organisms died off faster than they could reproduce.

10. When flies reproduce, they make maggots. Over a given time period, the maggots mature into adult flies in a process called metamorphosis.

11. While many organisms need a partner for reproduction, many do not. A bacterium, for example, simply makes copies of itself. Some flatworms actually tear themselves in half, and then each half regenerates so that there are two flatworms where there once was only one.

12. In general, the more dangerous an animal's life, the more offspring it will have. Rabbits, for example, are prey for all sorts of other animals. As a result, many die before they have a chance to reproduce. To make up for that, rabbits have many offspring. Most people, however, are able to stay alive and reproduce. As a result people don't have as many offspring as rabbits.

13. While some claim there is a problem with the population of people on the planet, it is just not true. The average number of babies being born to mothers throughout the world is decreasing. As a result, it is only a matter of time before the human population stops growing. In fact, while predictions such as these are often unreliable, the United Nations predicts that the world's population will stabilize by the year 2300.

14. Precocial offspring do not need nearly as much help from their parents when they are born as do altricial offspring. Puppies, for example, are altricial because they are born unable to see. The mother must take special care of them until their eyes open.

15. The smallest unit of life in creation is the cell. It is covered in an outer layer called a membrane, and a jellylike substance called cytoplasm fills the inside. In many of these units, small structures called organelles are suspended in the cytoplasm. They each have individual tasks they must accomplish. One of the most important structures is the nucleus, because it is where you will find most of the organism's DNA.

16. The three basic kinds of cells are animal cells (like those in your body), plant cells (found in plants), and cells with no nucleus (found in bacteria). The average animal cell is about 1 to 3 ten thousandths of an inch across. Also, it does not necessarily take a lot of them to make a living organism. Although many organisms are composed of trillions of cells, there are *billions and billions* of organisms that are composed of only one cell. Thus, a single cell can perform all of the functions of life and is therefore considered alive. Cells also reproduce, so if an animal cannot have offspring, it is still alive, because its cells reproduce.

ANSWERS TO THE SUMMARY OF MODULE #10

1. In the classification system we used, there are <u>five</u> kingdoms. Kingdom <u>Monera</u> contains all organisms that are composed of prokaryotic cells. Kingdom Protista contains those organisms that are composed of only one <u>eukaryotic</u> cell as well as <u>algae</u>. Kingdom <u>Fungi</u> contains mostly the organisms that feed on dead organisms. Kingdom Plantae is composed of organisms that are made of many eukaryotic cells and <u>produce their own food</u>. Kingdom <u>Animalia</u> contains organisms made of many eukaryotic cells and eat other (usually living) organisms.

2. If an organism is made of a single prokaryotic cell, it is a part of kingdom <u>Monera</u>, because all organisms made of prokaryotic cells are a part of that kingdom. If an organism is made of <u>eukaryotic</u> cells, it can be in any of the other four kingdoms. To determine which one, you have to know other details. For example, if it makes its own food, it is either a part of kingdom <u>Plantae</u> or kingdom <u>Protista</u>. To distinguish between these two kingdoms, you look at how many cells it is made of and its structure. If it is made of only one eukaryotic cell and makes its own food, it is in kingdom <u>Protista</u>. If it is made of many eukaryotic cells, makes its own food, and has specialized parts like roots, stems, and leaves, it is a part of kingdom <u>Plantae</u>. If it is made of many eukaryotic cells, makes its own food, and doesn't have roots, stems, and leaves, it is a part of kingdom <u>Protista</u>. If an organism is made of eukaryotic cells and eats dead things, it is a part of kingdom <u>Fungi</u>. If it is made of many eukaryotic cells and eats (mostly) living organisms, it is a part of kingdom <u>Animalia</u>. If it is made of only one eukaryotic cell and eats other (mostly) living things, it is part of kingdom <u>Protista</u>.

3. The members of kingdom Monera are often called <u>bacteria</u>. Many of these organisms can survive in habitats that are <u>deadly</u> to other organisms, and it is impossible to see them without the aid of a <u>microscope</u>. Some are <u>pathogenic</u>, which means they cause disease. Others, however, are actually helpful, such as those that live in our intestines and provide us with vitamin <u>K</u>. They require <u>water</u> to survive. Thus, they cannot survive and reproduce in <u>dehydrated</u> food. In addition, as your experiment showed, the presence of <u>salt</u> or <u>vinegar</u> can reduce the amount in food, as can <u>cold</u> temperatures. Since they can float on the dust particles in the air, <u>covering</u> food can help reduce the amount in it.

4. Kingdom Protista is typically split into two groups: <u>protozoa</u> and <u>algae</u>. In general, <u>protozoa</u> are able to move on their own, while <u>algae</u> are not. While algae use <u>photosynthesis</u> to make their own food, they are not plants, because they do not have <u>roots</u>, <u>stems</u>, and <u>leaves</u>. Algae are the <u>most</u> important source of oxygen for the planet. Like bacteria, some members of this kingdom are <u>pathogenic</u>, which means they can cause disease. Like bacteria, members of kingdom Protista need <u>water</u> to live.

5. Members of kingdom <u>Fungi</u> are called decomposers. They <u>recycle</u> dead matter so it can be used by the organisms in creation. While most are composed of <u>many</u> cells, there are some composed of only one cell. <u>Yeast</u> is an example of a single-celled fungus. The main body of a fungus is called the <u>mycelium</u>. It is often unseen, existing below the surface of its habitat. As a result, when you see a mushroom growing out of the ground, you are seeing only a <u>small portion</u> of the actual fungus.

6. Members of kingdom <u>Plantae</u> are made of several eukaryotic cells, use photosynthesis to produce their own food, and have specialized structures. The <u>roots</u> absorb water and nutrients from the soil, while the <u>stems</u> help transport the nutrients and water to the <u>leaves</u>. You can often grow a complete plant from just a portion of another plant, in a process called <u>vegetative</u> reproduction.

7. Plants cells are noticeably <u>different</u> from animal cells in several specific ways. They are usually more <u>square</u> than animal cells. Also, they have a <u>cell wall</u> that surrounds the outside of the membrane. They also have a <u>central vacuole</u> that fills with water, pushing the organelles against the <u>cell wall</u>. This causes pressure in the cell, called <u>turgor pressure</u>. This pressure helps a plant to <u>stand straight up</u>. A wilted plant may not be <u>dead</u>. It may just need water to reestablish its <u>turgor pressure</u>.

8. Kingdom <u>Animalia</u> contains those organisms that are made of many eukaryotic cells and eat other (mostly) living organisms. It contains <u>most</u> of the organisms with which people are familiar. While we normally think of cats, dogs, elephants, and the like as a part of this kingdom, there are <u>microscopic</u> members of this kingdom as well, such as the cyclops shown in the book.

9. People are a part of kingdom <u>Animalia</u>. This does not mean people are animals. It just means that from the standpoint of our cells and how our bodies are made, we have many things in <u>common</u> with animals. People, however, are unique in kingdom <u>Animalia</u>, as we are made in the image of <u>God</u>.

ANSWERS TO THE SUMMARY OF MODULE #11

1. The body's superstructure is composed of three units: the skeleton, the muscles, and the skin. The skeleton supports the body, and some of its bones are specifically designed to protect vital organs. In addition, a substance inside your bones, the red bone marrow, produces the cells that are in your blood.

2. In order to help you move, your body has about 640 different skeletal muscles. In addition to those muscles, there are also smooth muscles, which control the movements necessary for your body's internal organs and blood vessels to function. Finally, there is special muscle called cardiac muscle that is found in the heart. Under the microscope, smooth muscles appear smooth and unstriped, while skeletal muscles appear rough and striped. Skeletal muscles are voluntary (they are operated by conscious thought), while smooth muscles are involuntary (they are operated unconsciously by the brain). Cardiac muscle is also involuntary.

3. Your skin protects your body by preventing certain substances from getting inside. Also, it helps to sense the outside world. Skin cells harden through a process called keratinization. This process forms your hair, nails, and the outer layer of your skin.

4. Bones are as strong as steel but as light as aluminum. To this day, applied scientists cannot come up with any material that has this amazing mix of characteristics. Bones are most certainly alive, because they are composed of cells. The cells are surrounded by a substance called the bone matrix, which is composed principally of two things: collagen and minerals.

5. Collagen is a flexible, thread-like substance that belongs to a class of chemicals known as proteins. The minerals in bones are rigid, hard chemicals that contain calcium. Collagen and minerals work together to make your bones both strong and flexible. There are two main types of bone tissue – spongy bone and compact bone. The main difference between the two is how the minerals and collagen are packed together. In compact bone tissue, they are packed together tightly, forming a hard, tough structure that can withstand strong shocks. In spongy bone tissue, there are open spaces in the network of solid bone. This makes spongy bone lighter than compact bone.

6. A bone is surrounded by an outer sheath of tissue called the periosteum, which contains blood vessels that supply nutrients to the bones. It also contains nerves that send pain signals to your brain if the bone is damaged. Because bone is composed of living tissue, it continually changes to meet your body's needs. Not only do your bones change to meet your body's needs, they also grow as you grow.

7. The sum total of all bones in the body is called the endoskeleton. The vertebral column is often called the backbone, and members of kingdom Animalia that have one are called vertebrates. Members of kingdom Animalia that do not have one are called invertebrates.

8. The human endoskeleton can be split into two major sections: the axial skeleton (which supports and protects the head, neck, and trunk) and the appendicular skeleton (which attaches to the axial skeleton and has the limbs attached to it). In addition to bone, the human endoskeleton has cartilage, which is more flexible that bone.

9. Some members of kingdom Animalia have an exoskeleton, which supports and protects the creature but is on the outside of the body. Animals with an exoskeleton are called arthropods.

10. Skeletal muscles are attached to the skeleton by <u>tendons</u>. They can move the skeleton because the skeleton has joints. <u>Hinge joints</u> can be found at the elbow and the knee. They allow up and down (or left and right) motion, but that is all. <u>Ball-and-socket joints</u> are found at the hips and shoulders, and they allow for a wide range of motion. Your ankle is an example of a <u>saddle joint,</u> which allows a range of motion more or less in between the two joints previously mentioned. <u>Washer joints</u> exist only in your backbone and allow the smallest range of motion. In general, the <u>wider</u> the range of motion, the less stable the joint is.

11. The bones that make up a joint are covered in <u>articular cartilage</u>, which allows them to rub against each other without damage. In addition, the cartilage acts as a <u>shock absorber</u> so jarring movements do not destroy the joints. The bones in a joint are held together by strips of tissue called <u>ligaments</u>, which are like strips of stiff elastic that go from one bone to the other. Most joints are also surrounded by a "bag" called the <u>articular capsule</u>. Certain cells in this "bag" produce a fluid, called <u>synovial fluid</u>, which lubricates the joint. If a joint has such a "bag," it is a <u>synovial</u> joint.

12. Skeletal muscles work in groups of <u>two or more</u> to move the skeleton at the joints. These groups contain muscles that <u>contract</u> and <u>relax</u> to produce the motion. For example, to bend your arm at the elbow, you use the biceps brachii, which <u>contracts</u> to move your forearm at the elbow. Its partner is the triceps brachii, which <u>relaxes</u> and is passively stretched out while the biceps brachii <u>contracts</u>. When you straighten your arm out again, the biceps brachii <u>relaxes</u>, and the triceps brachii <u>contracts</u>.

13. Most members of kingdom Monera have <u>flagella</u>, which they use to move through the water (or other liquid) they inhabit. Some protozoa use tiny hairs called <u>cilia</u> that beat back and forth, acting like little oars that row the organism through the water. Plants also move. When they grow towards the light, it is called <u>phototropism</u>. When they move so that they always grow upward, it is called <u>gravitropism</u>.

14. Your skin is composed of two basic layers: the <u>epidermis</u> (which is the outside layer), and the <u>dermis</u> (which is found underneath). Below the dermis lies the <u>hypodermis</u>, which is composed mostly of fat and is not technically considered part of the skin.

15. Your epidermis is composed of a thick layer of <u>dead</u> cells that have been keratinized. This layer lies on top of a thin layer of <u>living</u> cells. A good fraction of the dust you find in your home is composed of the <u>dead</u> epidermal cells that have fallen from your family's (and your pets') skin.

16. Almost all of your skin produces <u>hair</u>, which is made from keratinized cells, but the keratin is harder in these cells than in the skin's cells. The cells that are keratinized come from a structure known as a <u>hair</u> follicle, which is like a tiny "pit" of epidermis. The lowest part of this structure, called the <u>matrix,</u> is the source of the cells that are keratinized. <u>Sebaceous glands</u> are connected to the <u>hair</u> follicle in the dermis. They produce oil that softens the <u>hair</u> and the <u>skin</u>.

17. <u>Sweat</u> is produced in your sweat glands, travels up through the dermis in the <u>sweat duct</u>, and then pours out onto your skin through your <u>sweat pores</u>. Sweating serves at least two purposes: it <u>cools</u> your skin and helps <u>feed</u> bacteria and fungi that live on your skin.

18. Hair has two functions: <u>insulation</u> and <u>sensation</u>. Animals with hair are <u>mammals</u>, while animals with feathers are <u>birds</u>. <u>Fish</u> and <u>reptiles</u> have scales. <u>Amphibians</u> breathe through their skin.

19. Your skin produces <u>food</u> for certain bacteria and fungi that help fight off <u>pathogenic</u> organisms. This situation, commonly found in creation, is an example of <u>symbiosis</u>.

ANSWERS TO THE SUMMARY OF MODULE #12

1. The energy in most living organisms originates in the sun. Organisms that produce their own food are called producers, and most get their energy directly from the sun. Organisms that don't get energy *directly* from the sun usually get it indirectly, by eating other organisms. They are called consumers. Organisms that recycle dead matter back into creation are called decomposers.

2. If a consumer eats only producers, it is called an herbivore. If it eats only other consumers, it is called a carnivore. If it eats both producers and consumers, it is an omnivore.

3. There are some living organisms that do not get their energy from the sun. For example, at the bottom of the ocean, hydrothermal vents spew forth superheated water that is rich in various chemicals. Some bacteria use those chemicals and the heat to make their own food, which makes them producers. These bacteria, and the consumers that eat them, do not get their energy from the sun.

4. Food is converted to energy via the process of combustion, which requires oxygen. It produces energy, carbon dioxide, and water.

5. There are only three things that your body can burn effectively: carbohydrates, fats, and proteins. Collectively, these three types of chemicals are called macronutrients, because we must eat a lot of them every day. If given a choice, your body would rather burn carbohydrates. If it can't burn those, it will resort to burning fats. If it can't get either of those, it will burn proteins.

6. Simple carbohydrates, like glucose, are called monosaccharides. When two of these simple carbohydrates link together, they form a disaccharide, such as table sugar (sucrose). If many simple carbohydrates link together, they form a polysaccharide, such as starch.

7. There are many types of fats, but they can all be put into one of two classes: saturated fats or unsaturated fats. Generally speaking, saturated fats are solid at room temperature, while unsaturated fats tend to be liquid at room temperature. Fat is absolutely essential for a healthy body. It insulates the body so the body can stay at its proper internal temperature. Many organs have a layer of fat for cushion as well. Also, much of the fat in your body serves as a great storehouse of energy in case you are unable to eat for an extended amount of time. Finally, there are certain vitamins that can only be stored in your body's fat reserves. Your body can produce most of the fats it needs from excess carbohydrates and proteins, but there are certain fats, called essential fats, that your body cannot produce. You must get those fats from your food.

8. Your body burns proteins only if you have too many of them, because they are essential to many other chemical processes that occur in your body. Proteins are formed when smaller chemicals, called amino acids, link together in long chains. Nearly every chemical reaction that occurs in the body is affected by proteins. The information stored in DNA is used to tell the cells in your body how to make proteins. Your cells can actually manufacture 11 of the 20 amino acids they need for the proteins they are required to make. However, there are nine amino acids, called essential amino acids, which your cells cannot manufacture. In order to get these amino acids, you must eat proteins that contain them. The best source of these essential amino acids is meat, milk, fish, and eggs.

9. People are endothermic, which means we use energy for the purpose of keeping our internal temperature relatively constant. In other words, we are "warm-blooded." The vast majority of

organisms are <u>ectothermic</u>, which means that they do not have a means by which they can control their internal temperature. In other words, they are "<u>cold</u>-blooded." <u>Ectothermic</u> organisms require less food than <u>endothermic</u> organisms, because it takes *a lot* of energy to maintain a constant internal temperature.

10. A <u>calorie</u> is a unit used to measure energy. Since food provides you with macronutrients, and since macronutrients provide you with energy, one way to measure the macronutrient content in food is to measure the number of <u>calories</u> of energy the food gives you. If you eat significantly fewer <u>calories</u> than you use, you will lose weight. If you eat significantly more <u>calories</u> than you use, you will gain weight.

11. The total rate at which your body uses energy is called your <u>metabolic rate</u>. It has two factors. The first, called the <u>basal metabolic rate (BMR)</u>, is the rate at which your body burns energy just to perform the minimum functions that will keep you alive. The second is the amount of <u>activity</u> you engage in every day.

12. Endothermic animals have a <u>higher</u> BMR than ectothermic animals, because it takes a lot of energy to maintain a constant internal body temperature. Ectothermic animals cannot be <u>active</u> on very cold days, because the speed at which the chemical reactions occur in their bodies is reduced by the resulting cooler body temperature.

13. Two people could eat the same amount of food and engage in the same activities, but one could gain weight while the other loses weight. If this happens, you know that the one who gained weight has a <u>lower</u> BMR than the one who lost weight.

14. The <u>smaller</u> the mammal, the larger its normalized metabolic rate. This is because small mammals have a <u>larger</u> percentage of their total body exposed to the outside air. As a result, a small mammal loses <u>more</u> heat than a large mammal. This requires the smaller mammal to burn <u>more</u> food to make up for the lost heat.

15. Food is burned in most living organisms through a three-step process. The first step is called <u>glycolysis</u>. In this step, the monosaccharide glucose is broken into <u>two</u> parts. This results in a small release of <u>energy</u> and a little bit of <u>hydrogen</u>. The two parts of the glucose and the hydrogen are then sent to a particular organelle in the cell called the <u>mitochondrion</u>, which is often called the "powerhouse" of the cell. In this "powerhouse," the process continues with the second step, called the <u>Krebs cycle</u>. In this step, the two pieces of glucose react with <u>oxygen</u> to produce <u>carbon dioxide</u> and <u>hydrogen</u>. That results in a small release of <u>energy</u> as well. The hydrogen from <u>glycolysis</u> as well as the hydrogen released in the <u>Krebs cycle</u> go through the third step, called the <u>electron transport system</u>. In this step, <u>oxygen</u> combines with all that <u>hydrogen</u> to make <u>water</u>. This results in a large release of <u>energy</u>.

16. The combustion process in living organisms is amazingly <u>complex</u>, because it must provide energy in a <u>gentle</u> but efficient fashion.

ANSWERS TO THE SUMMARY OF MODULE #13

1. When you eat your food, your body must <u>break it down</u> in order to get the nutrients contained within. That process is called <u>digestion</u>. There are two distinct parts to the process: <u>physical digestion</u> (where the food is simply broken into small pieces) and <u>chemical digestion</u> (where the chemical nature of the food is changed).

2. Food enters your body through your <u>mouth</u>. It is cut, crushed, and ground into little pieces by your <u>teeth</u>. It is also moistened by saliva, which is produced in your <u>salivary glands</u>. Your <u>tongue</u> moves food around your mouth and provides most of the taste sensation you get when you eat. It molds the food into a soft lump called a <u>bolus</u>. The food then goes to your <u>pharynx</u>, passes into your esophagus, and ends up in your <u>stomach</u>. There, it is churned and mixed with juices that are made by the <u>stomach</u> lining. The food is gradually released it into your <u>small intestine</u>, where it is broken down chemically. Most of the micronutrients and macronutrients are absorbed by the bloodstream through the lining of your <u>small intestine</u>. The nutrient-filled blood then passes through your <u>liver</u>, which picks up many of those nutrients. Once the food is digested in your <u>small intestine</u>, what was not absorbed is sent to your <u>large intestine</u>. By the time food reaches this point, most of the micronutrients and macronutrients have been <u>removed</u>. Water is absorbed from the remains, turning them into <u>feces</u>, which are then sent to your <u>rectum</u> and expelled from your body through your <u>anus</u>.

3. There are parts of the digestive system that never actually come in contact with the food being digested. The sum of the parts of the digestive system through which food actually passes is often called the <u>alimentary canal</u> or the <u>digestive tract</u>. However, that doesn't represent the *entire* digestive system, as there are some <u>digestive organs</u> through which food never travels.

4. Your teeth are made of hard, bonelike material and are surrounded by soft, shock-absorbent <u>gingiva</u>, which most people call "gums." The <u>incisor</u> teeth are sharp and used to cut food; the <u>canine</u> teeth are used to tear food; and the <u>premolars</u> and <u>molars</u> are used to crush and grind food.

5. When you swallow, your <u>soft palate</u> rises, sealing off the nasal cavity. The bolus then moves into the <u>pharynx</u>, which is a passageway for two things: <u>air</u> and <u>food</u>. When you inhale, <u>air</u> passes through the pharynx to the larynx and then into the lungs. When you swallow, however, the larynx rises up. This motion causes a small flap of cartilage called the <u>epiglottis</u> to cover the larynx so food goes only into the <u>esophagus</u>.

6. In the stomach, the bolus is mixed with a liquid called <u>gastric juice</u>. The most important chemical in this mixture is <u>hydrochloric acid</u>, which is sometimes called <u>stomach acid</u>. This chemical is a powerful acid that activates digestive chemicals and kills <u>pathogenic</u> microscopic organisms that might have been eaten along with the food. In addition, it helps <u>dissolve</u> the food so it is easier to digest. Smooth muscles in the stomach relax and contract, churning the bolus with the gastric juice until it is turned into a liquid mush called <u>chyme</u>.

7. Chyme passes from the stomach to the small intestine in spurts that are controlled by a ring of muscles called the <u>pyloric sphincter</u>. In the small intestine, the chyme is mixed with several more digestive chemicals which come from the <u>gall bladder</u> and <u>pancreas</u>. Because some of these chemicals cannot work properly in the presence of acids, <u>bases</u> from the pancreas, gall bladder, and small intestine are mixed with the chyme as it enters the small intestine. This <u>neutralizes</u> the stomach acid that is still in the chyme before the chyme reaches the small intestine. Unlike many organs, the inside

wall of the small intestine is not <u>smooth</u>. Instead, it is covered with millions of projections called <u>intestinal villi</u>. These villi increase the amount of intestinal wall that comes in contact with the food, <u>speeding</u> the absorption process.

8. The large intestine is actually composed of three parts: the <u>cecum</u>, the <u>colon</u>, and the <u>rectum</u>. Chyme enters the <u>cecum</u> from the small intestine, and smooth muscles in the cecum push it into the <u>colon</u>. The main function of the <u>colon</u> is to absorb water that is in the chyme. Many <u>bacteria</u> live in your large intestine. They feed on the chyme as it travels through the large intestine and produce chemicals like <u>vitamins</u> that are useful to your body. Anything that makes it to the <u>rectum</u> becomes a part of the feces, which are expelled through the <u>anus</u>.

9. For years, evolutionists have called the appendix a <u>vestigial organ</u>, which means they thought it was a useless remnant of the process of evolution. Like most evolutionary ideas, however, this has been demonstrated to be <u>false</u>. We now know that the appendix is a safe haven for <u>beneficial</u> bacteria that allows them to survive in the event of an intestine-clearing illness. That way, once the illness has passed, they can <u>repopulate</u> the intestines quickly, helping you regain your health.

10. The liver makes <u>bile</u>, which is a mixture of chemicals that prepares fats for digestion. This mixture is sent to the <u>gall bladder</u> where it is concentrated to increase its strength. In addition, the liver converts glucose into <u>glycogen</u> for storage. It also breaks down <u>glycogen</u> when the body needs energy. It stores fats and can convert them to <u>glucose</u> if the body needs energy. It can also convert amino acids into <u>glucose</u> when the body needs energy. It is also a <u>detoxification</u> center, recycling or transforming potentially harmful chemicals into useful or at least harmless substances. It also <u>warms</u> the blood to help regulate body temperature.

11. The nutrients that your body needs in small amounts are called <u>micronutrients</u>. You do not get <u>energy</u> from them, but they often support your body's chemical processes, increasing your overall health. They are generally split into two categories: <u>vitamins</u> and <u>minerals</u>.

12. Vitamins are effective in extremely <u>small</u> amounts and act mainly as regulators of the chemical processes that occur in your body. They are classified as either <u>fat-soluble</u> (A, D, E, and K) or <u>water-soluble</u> (C and the B-group). Your body can make two vitamins: <u>vitamin D</u> (made through the skin's exposure to sunlight) and <u>vitamin K</u> (produced by bacteria in the large intestine).

13. <u>Vitamin A</u> is a component of the process that allows your eyes to detect light. It also maintains the cells that protect your body, such as skin cells. <u>Vitamin D</u> allows your body to more effectively absorb certain minerals, especially calcium, from your food. <u>Vitamin E</u> is an "antioxidant," which means it helps protect certain important chemicals in your body from being destroyed through the chemical process called "oxidation." In addition, it helps repair your DNA. The <u>B vitamins</u> help your body's metabolism. <u>Vitamin C</u> helps your body build all sorts of molecules it needs. It is also an antioxidant. <u>Vitamin K</u> is an important part of that blood-clotting process.

14. The <u>minerals</u> you need most are those that contain calcium, phosphorus, magnesium, potassium, sodium, chloride, sulfur, chromium, copper, fluoride, iodine, iron, selenium, and zinc.

15. Although the micronutrients are essential for good health, anything can become toxic if it builds to a high enough concentration. Thus, <u>too many</u> vitamins and minerals can be dangerous. The vitamins that you must be most concerned about are the <u>fat-soluble</u> ones, as they are stored in your body's fat reserves and can build up over time.

ANSWERS TO THE SUMMARY OF MODULE #14

1. The human circulatory system is composed primarily of the heart and the blood vessels. It transports oxygen and nutrients to all the tissues. It also and picks up waste from the tissues and transports them to organs that can get rid of them.

2. The human respiratory system, composed primarily of the lungs, allows the body to take in oxygen from the surrounding air and expel carbon dioxide.

3. Blood vessels are separated into three basic categories: veins (vessels that carry blood to the heart), arteries (vessels that carry blood away from the heart), and capillaries (thin-walled vessels that allow for the exchange of gases and nutrients).

4. The pulmonary trunk takes blood away from the heart and splits into two arteries that take the blood to the lungs. There, the blood gets rid of carbon dioxide and receives oxygen. Then, the pulmonary veins take the blood back to the heart. It then leaves the heart through the aorta. Eventually, the blood reaches the capillaries, where it gives oxygen to the cells and picks up carbon dioxide. The blood is then picked up by the veins to be brought back to the heart. Eventually, all of the blood is returned to the heart via large veins called the superior vena cava and the inferior vena cava.

5. Like birds and all other mammals, humans have a four-chambered heart. Deoxygenated blood enters the heart in the right atrium. When the right atrium receives a signal from the sinoatrial node, it contracts, pushing blood into the right ventricle. The atrium then relaxes to fill with blood again, and the ventricle contracts, pushing the deoxygenated blood out the right ventricle and into the pulmonary trunk, which sends the blood to the lungs. The newly-oxygenated blood is carried back to the heart through the pulmonary veins, which dump blood into the left atrium. The atrium then contracts, sending the blood into the left ventricle. The atrium then relaxes and the ventricle contracts, pushing the oxygenated blood into the aorta so it can travel to the rest of the body. The entire cycle of a heartbeat – the contraction of the two atria, the relaxation of the atria and the contraction of the ventricles, and the relaxation of the ventricles – is called the cardiac cycle.

6. In most veins, the blood is deoxygenated, and in most arteries, the blood is oxygenated. The blood vessels that carry blood to the lungs and back are exceptions to this general rule. The blood in the arteries that go from the heart to the lungs is deoxygenated, while the blood in the veins that go from the lungs to the heart is oxygenated.

7. Blood is an incredibly complex mixture of chemicals and cells. The cells are produced in the bone marrow. More than half of any given sample of blood is made up of blood plasma. Suspended within this liquid are three main types of cells: red blood cells, white blood cells, and blood platelets. Red blood cells transport oxygen from the lungs to the tissues. They give the blood its overall red color, because they contain a protein called hemoglobin, which is red. White blood cells are responsible for protecting the body from agents of disease. Blood platelets are not true cells; rather, they are pieces of a kind of white blood cell. They aid in the process of blood clotting, which keeps you from bleeding to death when you are cut.

8. The respiratory system controls how you breathe. Air travels either through the nasal cavity or the oral cavity into the pharynx. From there, it travels into the trachea. The air eventually reaches a branch that marks the beginning of the lungs' bronchial tube system. A little more than half of the air

travels through the right branch into the <u>right lung</u>, and the rest travels through the left branch into the <u>left lung</u>. The right branch is called the <u>right primary bronchus</u>, and the left branch is called the <u>left primary bronchus</u>. These two primary tubes branch into smaller and smaller <u>bronchial</u> tubes. The tubes get smaller and smaller until they are tiny tubes called <u>bronchioles</u>. At the end of these tiny tubes, there are little sacs called <u>alveoli</u>, which is where the exchange of oxygen and carbon dioxide takes place.

9. The nasal cavity is lined with a sticky substance called <u>mucus</u>, which is designed to trap particles and keep them from reaching the lungs. When particles are trapped by the <u>mucus</u> of your nasal cavity, they are pushed towards the front of the nose by tiny hairs called <u>cilia</u>, where they will be blown out or sneezed out.

10. The process of breathing is controlled by a few skeletal muscles, the most important of which is the <u>diaphragm</u>. When this muscle contracts, it pushes down on the nearby organs, pulling your lungs down with them. This causes your lungs to <u>expand</u>, which sucks air into them. When the muscle relaxes, those same organs push up on the lungs, making them <u>smaller</u>. This forces air <u>out</u> of your lungs, and you exhale. When you exhale, air passes through the <u>larynx</u>, which is often called your voice box. It is called this because it contains your <u>vocal cords</u>, which are two thin folds of tissue that stretch across the sides of the larynx. As you exhale, air passes over these folds. When your <u>vocal cords</u> are relaxed, air passes over them <u>silently</u>. When your <u>vocal cords</u> are tightened, however, the folds move into the airway, and the air makes them <u>vibrate</u>, producing sound. Small amounts of air passing over your <u>vocal cords</u> produce soft sounds, while large amounts of air passing over your <u>vocal cords</u> produce loud sounds. The <u>pitch</u> of the sound is determined by how tight your <u>vocal cords</u> are while the air passes over them.

11. In animals, hearts vary from <u>one-chambered</u> to <u>four-chambered</u>, depending on the needs of the animal. Some animals, such as sponges, <u>don't</u> have a heart or a circulatory system. They have <u>mobile cells</u> that travel freely throughout their bodies, digesting food, transporting the nutrients to where they are needed, and exchanging oxygen for cell waste products. Animals such as fish have to extract the <u>oxygen</u> that is dissolved in the water. They use <u>gills</u> instead of lungs to accomplish this feat. Other animals, such as worms and amphibians, actually breathe through their <u>skin</u>! Insects have neither lungs nor gills. They have an intricate network of <u>tubes</u> that runs throughout the body, and air simply passes through them to be distributed throughout the body. Most plants have <u>xylem</u> (which transport water up from the roots to the rest of the plant) and <u>phloem</u> (which carry food from the leaves to the rest of the plant).

ANSWERS TO THE SUMMARY OF MODULE #15

1. The lymphatic system removes excess fluid from your body's tissues and returns it to the bloodstream. At the same time, it cleans the fluid of microorganisms and other contaminants that can cause health problems. The endocrine system produces hormones that regulate several of the chemical processes occurring in your body. The urinary system controls and regulates the balance of chemicals in your blood.

2. There is a vast network of lymph vessels that carry the watery fluid found between your body's cells. This clear fluid, called interstitial fluid, leaks out of capillaries and passes in and out of cells. When it is picked up by the lymph vessels, however, it is no longer called interstitial fluid. Instead, it is called lymph.

3. The end of a lymph vessel is closed, but the cells that form the vessel overlap to form "flaps" that allow interstitial fluid to leak inside. The lymph vessels are positioned in the body so that when certain muscles contract, the lymph vessels are squeezed, which causes a gentle flow of lymph through the lymphatic system. To keep the lymph flowing in the right direction, there are one-way "valves" that open when the lymph is flowing in the right direction and close to prevent it from flowing backwards.

4. Lymph nodes are the "filters" where the lymph is cleaned before it is returned to the blood. The spleen houses many white blood cells, which are there to grow and mature. These white blood cells also clean the blood that passes through. Interestingly enough, the spleen also acts as a "storehouse" for oxygen-rich blood. The tonsils and adenoids form a protective ring around the throat. They work together to produce and release antibodies that attack pathogens entering your body through your mouth or nose. The thymus gland has both lymphatic and endocrine functions. When you are young, your thymus gland is a place where certain white blood cells mature. They travel from the bone marrow to the thymus gland and actually "learn" how to do their job while they are there. This gland also releases a hormone called thymosin, which stimulates the development of the white blood cells known as T-cells.

5. A lymph node has several lymph vessels (called afferent lymph vessels) that bring in lymph, but only one vessel (the efferent lymph vessel) that takes lymph away. The cleaning power of lymph nodes comes from the lymphocytes, which is the name given to white blood cells found in the lymphatic system. There are several different kinds of lymphocytes, each of which performs different tasks. B-cells produce antibodies that attack specific disease-causing microorganisms. T-cells attack microorganisms directly. Macrophages scavenge the lymph, eating bacteria and other debris. When an infection is detected by the lymph nodes, the germinal centers of the lymph nodes release lymphocytes.

6. When B-cells produce antibodies, they also produce memory B-cells. These cells are configured to start producing the same antibody again the moment the same infection is detected. This gives the lymphatic system a memory, allowing it to react much more quickly if the body is attacked again.

7. The fact that the lymphatic system has a memory is the basis of a vaccine, which is one of medicine's greatest achievements. The overall goal of a vaccine is to trick the body into thinking it is infected. That way, it will make memory B-cells so it is ready to fight the real infection if it ever actually occurs.

8. There are two basic types of vaccines. The first type contains a <u>weakened</u> form of the pathogen itself. Since the pathogen is <u>weakened</u>, your body's immune system will destroy it before it can overtake your body, and it will produce the memory B-cells that will allow it to fight a full-strength infection if one ever occurs. The other type of vaccine contains a human-made <u>chemical</u> that makes your body react the same as if a certain pathogen has entered it. A vaccine is not a <u>cure</u> for a disease. Instead, it must be given <u>before</u> you are exposed to the disease so your body becomes ready to fight the disease if you are ever exposed to it. The act of giving someone a vaccine is often called <u>immunization</u>, because it makes a person immune to the disease. However, a small percentage of people will not <u>respond</u> to a vaccine, which means they will not become immune.

9. The body produces tears in the <u>lacrimal glands</u>, which are located on the top and side of each eyeball. Tears run from the <u>lacrimal glands</u> through tiny tubes called <u>tear ducts</u> and then flow across the eyes. Your body produces tears to <u>clean</u> and <u>lubricate</u> the eye, but you can also produce tears in response to strong <u>emotions</u>. The tears produced for the former reason are chemically quite <u>different</u> from the tears produced for the latter reason.

10. The urinary system is made up of the kidneys, ureters, bladder, and urethra. Each <u>kidney</u> is actually made up of about a million units called nephrons. The <u>renal artery</u> brings blood into the kidney. In order to be cleaned, blood flows to a nephron, where it is first <u>filtered</u>. Most of what is in the blood plasma (all nutrients, wastes, and water, but not the blood proteins) is temporarily dumped into the nephron. Then, as this fluid flows through the nephron, the cells that line the nephron <u>reabsorb</u> the proper amounts of nutrients and chemicals back into the blood. Excess chemicals are left behind in the nephron to become <u>urine</u>. The blood then leaves the kidney through the <u>renal vein</u> and travels back to the heart. Any water and chemicals that were not reabsorbed into the blood go from the nephrons into the <u>renal pelvis</u> and flow out of the kidney to the <u>ureter</u>. At this point, the mixture of water and chemicals is called <u>urine</u>. It travels through the <u>ureter</u> and is held in the <u>bladder</u>. Eventually, the bladder releases the <u>urine</u> it has stored, and it leaves the body through the <u>urethra</u>.

11. <u>Hormones</u> are released by endocrine glands that are scattered throughout the body. The <u>hypothalamus</u> is one of the main regulators of the endocrine system. It is part of the <u>brain</u>, and it influences a wide range of body functions. Its function in the endocrine system is to control the <u>pituitary gland</u>, which is often referred to as the "<u>master</u> endocrine gland." It is given this name because the hormones it makes and puts into the bloodstream control many other <u>endocrine glands</u> in the body.

12. The <u>thyroid gland</u> produces hormones that affect the basal metabolic rate. The <u>parathyroid glands</u> are tiny glands that are on the edges of the thyroid gland. Their main job is to regulate the level of <u>calcium</u> in the body. The <u>adrenal glands</u> release cortisol, which is part of the "fight or flight" response. They also produce the hormones epinephrine and norepinephrine, which are also released during times of <u>stress</u>. Although the <u>pancreas</u> has digestive functions, it is also an endocrine gland. Cells located within "islands" of the <u>pancreas</u> make insulin, a hormone that enables glucose to enter the cells so it can be burned.

13. Often, the endocrine system has hormones that work toward opposite goals. Because they work "against" each other, such hormones are often called <u>antagonists</u>. They don't really work against each other; however, they just work at different <u>times</u>.

ANSWERS TO THE SUMMARY OF MODULE #16

1. The work of the nervous system occurs in <u>cells</u> called neurons. They communicate with other neurons through <u>neural pathways</u>. Signals pass from one neuron to another at a <u>synapse</u>. Neurons cannot function properly without the help of <u>neuroglial cells</u>, which are often called <u>neuroglia</u>.

2. The control center of a neuron is its nucleus, which is housed in the <u>cell body</u>. Signals are sent to the <u>cell body</u> along fibers called <u>dendrites</u>, and they are sent to other neurons along a single fiber called the <u>axon</u>. The axon is often covered by a fatty insulator called a <u>myelin sheath</u>, which protects the <u>axon</u> and speeds up the rate at which an electrical signal can travel down it.

3. At a synapse, the end of the axon is full of tiny sacs that contain chemicals we call <u>neurotransmitters</u>. When a signal comes to the end of an axon, these chemicals are <u>released</u> from their sacs. They travel across the synapse, chemically interacting with <u>receptors</u> on the cell at the other side of the tiny gap. This generates a new <u>signal</u>, which can then be used by the receiving cell.

4. The nervous system is split into two components: the <u>central nervous system</u> (CNS) and the <u>peripheral nervous system</u> (PNS). The CNS is composed of both the <u>brain</u> and the <u>spinal cord</u>. The PNS contains all the neurons that are involved in <u>receiving</u> information and <u>sending</u> it on to the spinal cord and brain. It also contains the neurons responsible for <u>transmitting</u> signals from the CNS to the various parts of your body that need to be controlled.

5. The spinal cord is protected by the <u>vertebral column</u>. The vertebrae that make up the <u>vertebral column</u> have a hole in their center, which lines up with the holes in the other vertebrae. This forms a <u>tunnel</u> through which the spinal cord passes. The brain sits on "shelves" inside the <u>skull</u>, which protects it from harm. In addition, the brain floats in liquid called <u>cerebrospinal fluid</u>. Although this liquid provides chemicals to the brain, its main function is <u>protection</u>. In addition, <u>cerebrospinal fluid</u> serves as a way for doctors to diagnose certain problems related to the nervous system. If there is something wrong in the nervous system, the <u>cerebrospinal fluid</u> balance of chemicals will most likely not be correct, so examining the liquid can be the first step in understanding a patient's problem.

6. The brain is divided into halves called <u>hemispheres</u>. The right side sends signals to the PNS on the <u>left</u> side of the body, and the left side sends signals to the PNS on the <u>right</u> side of the body. The folded tissue that surrounds the outside of the brain is called the <u>cerebrum</u>, which deals with what are often called "higher-level" brain functions. The <u>corpus callosum</u> is composed of axons running crosswise between the hemispheres, which allow the two hemispheres of the brain to exchange information. The <u>cerebellum</u> has a lot of functions, mostly oriented around muscle movements. The <u>brain stem</u> is right next to the cerebellum, and it controls the more basic functions of the human body. The <u>hypothalamus</u> not only controls the pituitary gland, but it also regulates thirst, hunger, and body temperature. It also helps initiate the "fight or flight" response. While some claim people use only 10% of the brain, it is <u>not</u> true.

7. Gray matter is composed almost exclusively of the <u>cell bodies</u> of neurons. White matter is composed mostly of the <u>axons</u> of the neurons in the gray matter.

8. The <u>left</u> side of the cerebrum tends to be responsible for speaking, logic, and math. The <u>right</u> side is more involved with spatial relationships, recognition, and music. In addition, one side of the body tends to be <u>dominant</u> over the other in a given individual. For example, a person will generally write

with only one hand, because that's his <u>dominant</u> hand. In the majority of people, the right side of the body is <u>dominant</u>.

9. The blood-brain barrier <u>insulates</u> the brain from the blood. It is important because many of the chemicals in your blood are <u>toxic</u> to your brain cells. The blood-brain barrier selectively <u>transports</u> "good" chemicals into the brain and leaves the "bad" chemicals in the capillaries, away from the brain.

10. The <u>PNS</u> is made up of those nerves that run off of the CNS. A nerve is made up of <u>dendrites</u> and <u>axons</u>, not the cell bodies of neurons. The cell bodies of neurons typically cluster together in groups called <u>ganglia</u>.

11. The PNS is composed of three main divisions: the <u>autonomic nervous system</u>, the <u>sensory nervous system</u>, and the <u>somatic motor nervous system</u>. The <u>autonomic nervous system</u> carries instructions from the CNS to the body's smooth muscles, cardiac muscle, and glands. The <u>sensory nervous system</u> carries information from the body's receptors to the CNS. The <u>somatic motor nervous system</u> carries instructions from the CNS to the skeletal muscles.

12. The autonomic nervous system is composed of two parts: the <u>sympathetic division</u> and the <u>parasympathetic division</u>. These two divisions <u>counterbalance</u> each other in many ways. The <u>sympathetic division</u> increases the rate and strength of the heartbeat and raises the blood pressure. It also stimulates the liver to release more glucose in the blood, producing quick energy for the "fight or flight" response. The <u>parasympathetic system</u>, on the other hand, slows the heart rate, which lowers the blood pressure. In addition, it takes care of certain "housekeeping" activities such as causing the stomach to churn while it is digesting a meal.

13. The sense of taste is called the <u>gustatory</u> sense. Your tongue has holes called <u>taste pores</u> that lead to clusters of cells called <u>taste buds</u>. These cells have tiny "hairs" that are sensitive to certain <u>chemicals</u>. When those <u>chemicals</u> are detected, signals are sent to the brain, generating a <u>taste</u> sensation. Scientists think there are only <u>five</u> basic taste sensations: <u>sweet</u>, <u>salty</u>, <u>sour</u>, <u>bitter</u>, and <u>umami</u>. It is thought that all tastes are a <u>combination</u> of these five sensations.

14. The roof of your nasal cavity houses mucus-covered tissue called the <u>olfactory</u> epithelium. This tissue has cells, called <u>olfactory sensory cells</u>, that have long "hairs" that stick into the mucus. When chemicals in the air dissolve in the mucus, they interact with the "hairs" of the <u>olfactory sensory cells</u>. This causes the cells to send signals to your brain, which gives you the impression you call <u>smell</u>. This sense affects the <u>gustatory</u> sense.

15. When light strikes the eye, it first passes through the <u>cornea</u>. It then passes through a clear liquid called the aqueous humor and then through the <u>pupil</u>. It then passes through the <u>lens</u>, which focuses the image onto the <u>retina</u>. The lens can focus light because it changes <u>shape</u> based on the actions of the <u>ciliary muscle</u>. The <u>retina</u> is filled with light-sensing cells called <u>rods</u> and <u>cones</u>.

16. Your sense of touch is all over your body, so it is called a <u>somatic sense</u>. The number of touch <u>receptors</u> in an area of skin determines how sensitive that part of your body is to touch.

17. The <u>external</u> ear acts as a "funnel" to send vibrations in the air down the <u>ear canal</u> until they reach the <u>ear drum</u>. Vibrations in the air cause the ear drum to vibrate, which cause tiny bones called <u>ear ossicles</u> to move back and forth. This movement in turn vibrates fluid within the snail-shaped <u>cochlea</u>.

Cells in the fluid of the <u>cochlea</u> pick up the vibration and convert it into an electric signal that is sent to the brain, and the brain interprets the signals as <u>sound</u>.

TEST FOR MODULE #1

1. Define the following terms:

a. Science
b. Papyrus
c. Spontaneous generation

MATCHING
Match the person on the left with the proper description on the right

2. James Clerk Maxwell

3. Thales

4. Galileo

5. Ptolemy

6. Einstein

7. Lavoisier

8. Darwin

9. Democritus

10. Grosseteste

11. Copernicus

12. Aristotle

13. Newton

14. Mendel

15. Joule

16. Dalton

17. Bohr

a. Destroyed the idea of the immutability of species

b. Demonstrated the First Law of Thermodynamics

c. He is best known for his model of the atom. It was named after him, and it revealed many of the atom's mysteries.

d. Discovered the Law of Mass Conservation

e. One of the first scientists

f. Ancient Greek scientist who believed in atoms

g. He developed the first detailed atomic theory and became known as the founder of modern atomic theory.

h. Determined how traits are passed on during reproduction

i. Considered the first modern scientist

j. Had two theories of relativity and was big in quantum mechanics

k. Founder of modern physics

l. Championed the idea of spontaneous generation and is responsible for it being believed for so long

m. Proposed heliocentric system

n. Proposed geocentric system

o. One of the greatest scientists of all time, he laid down the laws of motion, developed the law of universal gravity, and invented calculus

p. Collected much data in favor of the heliocentric system but was forced to recant belief in it

18. What lesson can we learn from the fact that scientific progress stalled during the Dark Ages?

19. What caused scientific progress to move forward again towards the end of the Dark Ages?

20. What lesson can we learn from the fact that the idea of spontaneous generation was believed for so long, despite the evidence against it?

TEST FOR MODULE #2

1. Define the following terms:

a   Counter-example
b. Hypothesis
c. Theory
d. Scientific law

2. Put the following steps of the scientific method into their proper order:

a. Theory is now a law
b. Hypothesis is now a theory
c. Make observations
d. Perform experiments to confirm the hypothesis
e. Form a hypothesis
f. Perform many experiments over several years

3. What is wrong with the following statement?

All objects, regardless of their weight, fall at the same rate

Questions 4 through 7 refer to the following story:

A slightly eccentric student is standing by a pool counting his pennies. He drops a penny and notices that it sinks to the bottom of the pool. He decides that all solid objects sink in water. Thus, the student starts dropping objects into the pool. He drops all his coins, some rocks, books, a chair, and his shoes in the pool. All of them sink. He then proudly states that he has come up with a theory: All solid objects sink in water. Another student drops a cork into the pool and it floats. The eccentric student is crestfallen.

4. Did the eccentric student follow the scientific method?

5. If you answered "yes" to question 4, list the observation, hypothesis, and experiment designed to confirm the hypothesis. If you answered "no," explain why.

6. What did the other student provide to destroy the eccentric student's theory?

7. How is this story similar to the story about the theory that there are canals on Mars?

8. When scientists discovered high-temperature superconductors, it was quite surprising. Why?

9. What are the three limitations of science?

10. There is a lot of interest in how life originated on this planet. Can such a subject be studied by science?

TEST FOR MODULE #3

1. Define the following terms:

a. Experimental variable
b. Control (of an experiment)
c. Blind experiments
d. Double-blind experiments

2. Why is it important to analyze an experiment for experimental variables?

Questions 3 through 6 refer to the following story:

A consumer lab decides to test the claims of certain automatic dishwasher additives. Three brands claim to reduce the amount of water spots on dishes when you add them to your dishwasher in addition to your normal dishwashing liquid. The lab does a load of dirty glasses with a standard brand of dishwashing detergent with no additives. The lab then does three more loads, each time adding a different brand of additive to the same standard detergent. Each load of glasses is different, but they are all done in the same dishwasher, one after the other. Each load is inspected by eye, and the cleanest load is determined.

3. Which of the following are experimental variables?

a. The dishwasher used
b. The additive used
c. The glasses used
d. The detergent used
e. The cleanliness of the dishwasher before each load is washed

4. From which of the experimental variables you identified above will the lab learn something?

5. Which of the experimental variables you identified above should be reduced or eliminated?

6. Is the data collected from this experiment subjective or objective?

7. Two things are floating on the surface of a sink full of water: a cork and a metal paper clip. What happens to the two items when dishwashing soap is put in the water?

8. A certain nutritional bar is supposed to give runners an extra "boost," allowing them to run without getting as tired as they normally do. To test the effectiveness of this nutritional bar, a scientist gets a group of volunteers together. She decides to give half of the volunteers the nutritional bar and the other half a regular granola bar. She then makes them run 3 miles and tell her whether they feel more or less tired than they usually do after a 3-mile run. Should this study be done as a single-blind experiment, a double-blind experiment, or neither?

Questions 9 through 11 refer to the following story:

To investigate how well lead protects against radioactivity, a scientist puts a radiation detector near a radioactive source. He then puts varying amounts of lead in between the radiation detector and the radiation source and measures how much radiation reaches the detector. His results are as follows:

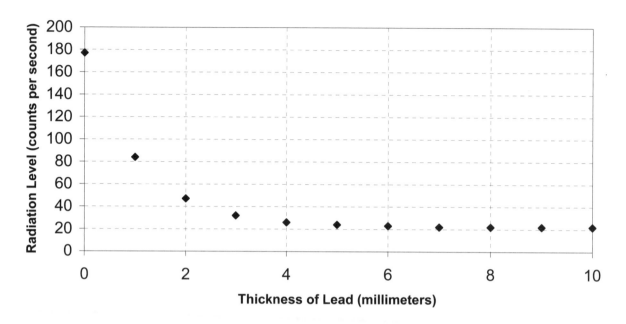

9. What is the radiation level that the detector sees when no lead is used to block the radiation?

10. If the scientist wants to reduce the radiation level to about 50 counts per second, how many millimeters of lead should he use?

11. You are designing shielding for a lab. Your budget is tight, however. You want as much protection as possible, but you don't want to buy any more lead than is necessary. How many millimeters of lead should you use in your shielding?

TEST FOR MODULE #4

1. Define the following terms:

a. Simple machine
b. Force
c. Mechanical advantage
d. Diameter
e. Circumference

2. Which of the following experiments would be considered applied science experiments?

a. An experiment to find a better gasoline mixture for automobiles
b. An experiment to determine how to reduce pollutants produced by automobiles
c. An experiment to determine the level of pollutants in the air
d. An experiment to find the source of pollutants in the air

3. Which of the following are examples of technology?

a. A detailed description of the eating habits of bears
b. A detailed population analysis of the bears in an area
c. A gun for hunting bears
d. A spray that, when sprayed around a campsite, keeps bears away

4. If you cannot generate enough force to lift a rock with a first-class lever the way you have it set up, should you change the setup so that the fulcrum is closer to or farther from the rock?

5. Scissors are an example of two levers put together. To which class do the levers belong?

6. If a third-class lever has a mechanical advantage of 5, what does that mean?

7. What is the mechanical advantage of a wheel and axle system if it has a wheel with a diameter of 20 inches and an axle with a diameter of 2 inches?

8. If you turn the wheel of a wheel and axle, what does the mechanical advantage do for you? What is the drawback that accompanies the mechanical advantage?

9. A block and tackle uses two pulleys. What is the mechanical advantage?

10. In the diagram below, is the simple machine an inclined plane or a wedge?

11. If the slope of the machine in problem #10 is 10 inches and the height is 2 inches, what is the mechanical advantage?

12. What is the mechanical advantage of a screw that has a pitch of 0.05 inches and a head diameter of 0.1 inches?

13. If you turn the screw in problem #12 with a screwdriver that has a diameter is 1 inch, what is the mechanical advantage?

TEST FOR MODULE #5

1. Define the following terms:

a. Life science
b. Archaeology
c. Geology
d. Known age
e. Dendrochronology
f. The Principle of Superposition

2. When testing for the historical accuracy of a document, which test evaluates whether or not the document we have today is the same document as the original?

3. When testing for the historical accuracy of a document, which test evaluates whether or not the document contradicts itself?

4. When testing for the historical accuracy of a document, which test evaluates whether or not the document contradicts other known historical or archaeological facts?

5. Although the genealogies of Jesus in Luke 3 and Matthew 1 seem to contradict each other, they really do not. Why? (You can use a Bible if you like.)

6. Of the three tests for a document of history, which two does the Bible pass better than any other document of its time?

7. Why must we apply Aristotle's dictum when using the internal test?

8. An archeologist finds a preserved log that was used to build an old shack. The rings of the log are clearly visible. In order to determine the age of the shack by dendrochronology, what must the archaeologist find in those rings?

9. Which type of age is more certain: a known age or an absolute age?

10. What kind of ages does dendrochronology provide, known ages or absolute ages?

11. If an artifact has neither a known age nor an absolute age, what principle might be used to determine its age relative to something else?

12. The principle mentioned in problem #11 makes a very important assumption. What is that assumption?

TEST FOR MODULE #6

1. Define the following terms:

a. Catastrophism
b. Uniformitarianism
c. Humus
d. Minerals
e. Weathering
f. Erosion
g. Unconformity

2. Which hypothesis (uniformitarianism or catastrophism) can allow for an earth that is just a few thousand years old?

3. A rock is formed and then later on is changed as a result of heat and pressure. What kind of rock is it now?

4. A layer of rock is laid down by water. What kind of rock is it?

5. A layer of rock is formed from lava that erupts from a volcano. What kind of rock is it?

6. An igneous rock is laced with iron. When exposed to water, the iron rusts and the rock crumbles. Is this chemical or physical weathering?

7. A river flows through 2 regions of the country. The first region has few plants, while the second region is covered with thick grass, flowers, and many trees. In which region do you expect the most erosion to occur?

8. A cavern has no groundwater seeping through its ceiling. Will you see stalactites in it? Will you see stalagmites in it?

(TEST CONTINUES ON THE NEXT PAGE)

Questions 9 through 12 refer to the side-on diagram of the Grand Canyon below:

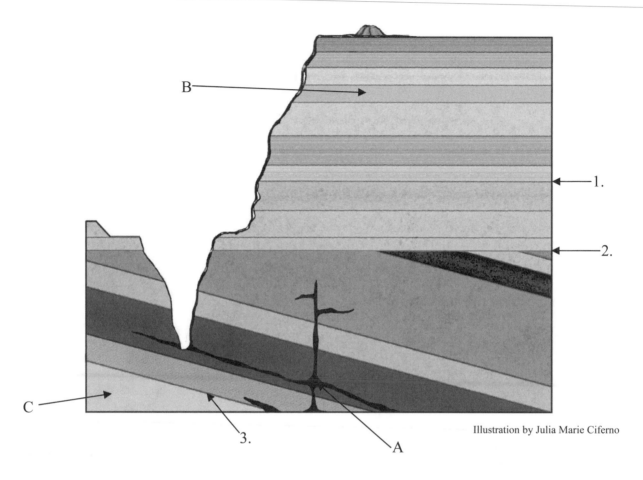

Illustration by Julia Marie Ciferno

9. Identify the type of rock (igneous, metamorphic, or sedimentary) pointed out by each *letter* in the figure.

10. Which *number* in the figure points to the angular unconformity?

11. Which *number* in the figure points to a disconformity?

12. Which *number* in the figure points to a nonconformity?

TEST FOR MODULE #7

1. Define the following terms:

a. Fossil
b. Petrifaction
c. Resin
d. Extinct

Questions 2 through 6 refer to the following fossil types:

Molds
Casts
Petrified remains
Carbonized remains
Creatures trapped in amber
Creatures trapped in ice

2. Which two kinds of fossils preserve only the shape and outer details of the creature?

3. Which two fossil types give the most detail about the original creature?

4. Which fossil type depends on one of the other fossil types being formed first?

5. Which fossil type requires mineral-rich water in order to form?

6. If a plant is fossilized, which fossil type will it most likely be?

7. What are the four general features of the fossil record?

8. A geologist believes that every layer in a series of stratified rock represents a different time period in earth's past. Is this geologist a uniformitarian or a catastrophist?

9. A geologist believes that most fossils were all laid down over a short time period in earth's past. Is this geologist a uniformitarian or a catastrophist?

10. Is it possible to believe that the earth is only a few thousand years old if you are a uniformitarian?

TEST FOR MODULE #8

1. Define the following terms:

a. Index fossils
b. Geological column
c. The Theory of Evolution

Problems 2 through 7 refer to the figure below, which illustrates two different geological formations a geologist is studying. The letters are just labels for the strata, and the symbols represent index fossils in each layer.

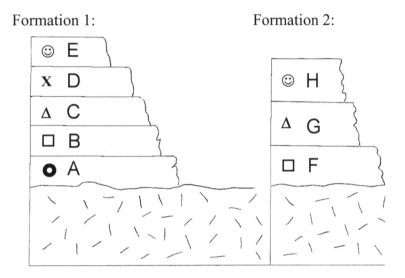

2. Using uniformitarian assumptions, identify the layer in formation 1 that corresponds to the same time period as the rock in layer F.

3. Using uniformitarian assumptions, identify the layer in formation 1 that corresponds to the same time period as the rock in layer G.

4. Using uniformitarian assumptions, identify the layer in formation 1 that corresponds to the same time period as the rock in layer H.

5. Considering both formations, which layer (A-H) would uniformitarians say is the very oldest layer of rock?

6. According to uniformitarian assumptions, which time periods represented by the layers in formation 1 are not in formation 2?

7. According to uniformitarians, which creatures came first: the ones whose fossils are represented by squares or the ones whose fossils are represented by smiley faces?

8. Why is the data from Mt. Saint Helens considered evidence for catastrophism?

9. Is the geological column a real thing?

10. Does the fossil record support the idea of evolution or the idea that God created each kind of plant and animal individually?

11. What is a paraconformity? Which viewpoint (uniformitarian or catastrophist) requires the existence of paraconformities?

12. Name two problems with uniformitarianism.

13. Name two problems with catastrophism.

14. Do fossils require a long time (like thousands or millions of years) to form?

TEST FOR MODULE #9

1. Define the following terms:

a. Atom
b. Molecule
c. Photosynthesis
d. Metabolism
e. Receptors
f. Cell

2. Here are two criteria for life:

*   All life forms contain DNA.
*   All life forms can sense changes in their surroundings and respond to those changes.

Which two are missing?

3. In the following drawing:

Illustration © Risteski Goce
Agency: www.shutterstock.com

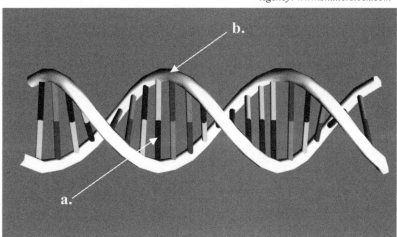

Which arrow is pointing to the backbone and which is pointing to a nucleotide base?

4. How is information stored in DNA?

5. One half of a portion of DNA has the following sequence of nucleotide bases:

thymine, cytosine, guanine, adenine, guanine, thymine

What is the sequence of nucleotide bases on the other half of this portion?

6. What two things (use their proper, chemical names) are produced by photosynthesis?

7. Fill in the blank: Metabolism produces energy, carbon dioxide and _____.

8. If a certain type of organism produces lots of offspring, what can you conclude about the level of danger that the organism experiences throughout the course of its life?

9. Even if a person never has children, that person still reproduces. How?

10. In the following picture:

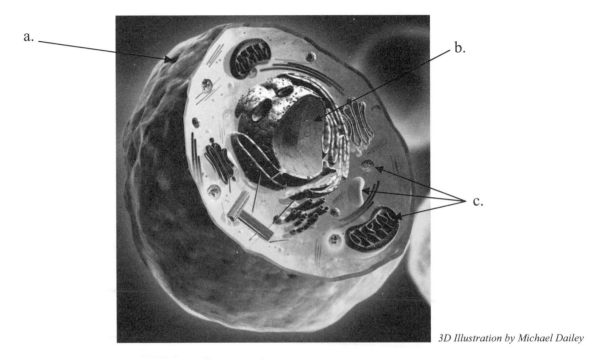

3D Illustration by Michael Dailey

Indicate what the arrows are pointing out.

TEST FOR MODULE #10

1. Define the following terms:

a. Prokaryotic cell
b. Eukaryotic cell
c. Pathogen
d. Decomposers
e. Vegetative reproduction

2. Which kingdom is missing from the following list?

Monera, Plantae, Fungi, Animalia

3. An organism is made of many eukaryotic cells and eats other, living organisms. To which kingdom does it belong?

4. An organism is composed of one prokaryotic cell and eats other, living organisms. To which kingdom does it belong?

5. An organism is made of many eukaryotic cells and makes its own food. It has roots, stems, and leaves. To which kingdom does it belong?

6. An organism is made up of one eukaryotic cell and eats only dead organisms. To which kingdom does it belong?

7. Three strips of bacon are left lying on a table for 2 hours. The first is left uncovered. The second is salted. The third is salted and covered with plastic wrap.

    a. Which strip of bacon will be most contaminated by bacteria?

    b. Which will be least contaminated by bacteria?

8. What is the main difference between algae and plants? Which kingdom contains algae, and which kingdom contains plants?

9. Are decomposers an essential part of creation?

10. A man walks out into his back yard and sees several mushrooms growing. He doesn't want mushrooms in his yard, so he picks all of the mushrooms, tearing them off at ground level. Afterwards, he sees no more mushrooms in his yard. The next day, he is horrified to see more. He picks all of them just as he did the day before. The next day, he sees more mushrooms again. Why do the mushrooms keep re-appearing?

11. When a plant wilts, does that automatically mean it is dead?

12. What two parts of a plant cell work together to make turgor pressure?

TEST FOR MODULE #11

1. Define the following terms:

a. Axial skeleton
b. Appendicular skeleton
c. Exoskeleton
d. Symbiosis

In questions 2 through 11, Match the word on the left with the phrase that best describes it on the right.

2. smooth muscles   a. Makes bones flexible

3. skeletal muscles   b. Process by which cells are hardened and die in order to make
                               hair, nails, and the outer layer of skin

4. keratinization   c. Make bones hard

5. bone marrow   d. Cushions the bones in a joint so they do not rub together
                               painfully

6. collagen   e. Involuntary muscles

7. minerals   f. Holds bones together in a joint

8. arthropods   g. Connects skeletal muscles to the skeleton

9. ligament   h. Animals with exoskeletons

10. cartilage   i. Voluntary muscles

11. tendon   j. Makes blood cells

12. Consider two joints. The first has a large range of motion while the second's range of motion is limited. Which, most likely, is the more stable joint?

13. A student experiments with 2 plants. The first one he waters but keeps in a dark closet. The second one gets watered and is also in a dark closet. The closet in which this plant is kept, however, has a window through which light can enter the closet. In three days, the first plant is dead and the second one has grown so that its leaves face the window. Which plant demonstrates phototropism?

14. A person's sweat glands are malfunctioning so that the person never sweats. Why is this person more likely to get sick than a person whose sweat glands are working?

15. Classify the following animals as mammal, reptile, amphibian, or bird.

    a. An ostrich – a creature that has feathers but cannot fly
    b. A brown bear with a thick, fur coat
    c. A salamander that breathes through its skin

TEST FOR MODULE #12

1. Define the following terms:

a. Producers
b. Consumers
c. Herbivore
d. Carnivore
e. Omnivore
f. Basal metabolic rate

2. Label each of the following as a consumer, producer, or decomposer:

a. yeast
b. ant
c. fly
d. corn stalk
e. rose bush

3. Fill in the blanks:

The process of combustion requires _____ and makes _____, _____, and _____.

In numbers 4 through 10, match the word on the left with the best description on the right:

4. carbohydrates   a. Usually the second macronutrient that is burned by the body

5. monosaccharides   b. The longest of the carbohydrates

6. mitochondrion   c. An organism that does not have a constant internal temperature

7. fats   d. Usually the last macronutrient that is burned by the body

8. proteins   e. Usually the first macronutrients burned by the body

9. polysaccharides   f. The powerhouse of the cell

10. ectothermic   g. The shortest of the carbohydrates

11. Why do endothermic organisms have a higher BMR than ectothermic organisms?

12. When a person doesn't eat enough protein or doesn't eat the right kinds of protein, what can the cells no longer do properly?

13. If you use lard or shortening (fats that are solid at room temperature) while you are cooking, are you using saturated fats or unsaturated fats?

TEST FOR MODULE #13

1. Define the following terms:

a. Digestion
b. Vitamin

2. Identify the organs pointed out by the arrows in the figure. Use the names at the left of the figure:

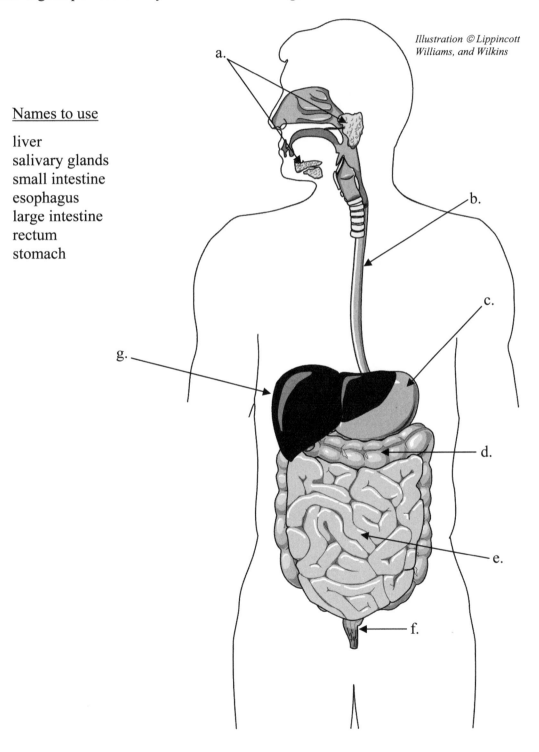

*Illustration © Lippincott
Williams, and Wilkins*

Names to use

liver
salivary glands
small intestine
esophagus
large intestine
rectum
stomach

3. In which organ does most of the absorption of nutrients occur?

4. In which organ is the bolus turned into chyme?

5. In which organ is the undigested food turned into feces?

6. Which organ produces the sodium bicarbonate that helps to neutralize the stomach acid in the chyme?

7. Which organ produces bile?

8. Which organs produce saliva?

9. Which organ moves the food around in the mouth to form the bolus?

10. What does the epiglottis do?

11. For the following vitamins:

vitamin D, vitamin C, B vitamins, vitamin E, vitamin K

a. Which are water-soluble?
b. Which can be absorbed by the body even if they are not in any food that is eaten?
c. Which are the most likely to build up to toxic levels if you take too many vitamin pills?

TEST FOR MODULE #14

1. Define the following terms:

a. Veins
b. Arteries
c. Capillaries

2. Starting with the vena cava dumping a sample of deoxygenated blood into the heart, name each chamber of the heart in the order in which that sample will pass through it.

3. From which chamber of the heart does deoxygenated blood leave on its way to the lungs?

4. If a sample of blood is oxygenated, did it most likely come from an artery or a vein? Can you be 100% certain?

5. What makes up the majority of the blood: red blood cells, white blood cells, platelets, or plasma?

6. What is blood clotting? What cells in the blood aid that process?

7. What blood cells contain hemoglobin?

8. What blood cells fight disease-causing organisms in the blood?

9. You are given a sample from a person's lungs. Looking at it under the microscope, you see many little round sacs that are covered in capillaries. What structures are you looking at?

10. When you *exhale* air, does the air pass through the trachea or the pharynx first?

11. You are listening to a man and a woman singing. The man is singing loudly with a very low pitch. The woman is singing much more softly but at a high pitch. Which singer is passing more air over his or her vocal cords?

12. Suppose you are looking through a microscope at a sample of living tissue. You watch the blood vessels and notice that the blood is going from oxygenated to deoxygenated as it passes through the vessels. What kind of blood vessels are you looking at: arteries, veins, or capillaries?

13. What are xylem? What is their purpose?

14. Where are blood cells produced?

TEST FOR MODULE #15

1. Define the following terms:

a. Gland
b. Vaccine
c. Hormone

2. Which of the systems in the human body is chiefly responsible for regulating water and chemical levels in the fluids of the body?

3. Which of the systems in the human body is chiefly responsible for fighting disease?

4. In what structures can you find most of the infection-fighting cells of the lymphatic system?

5. What pumps lymph through lymph vessels?

6. What cells in the lymphatic system produce antibodies?

7. What cells in the lymphatic system give vaccines the ability to make people mostly immune from certain diseases?

8. A doctor has two medicines that fight the same disease. The first medicine is given if the patient has the disease. The second is given to patients who have never had the disease. Which of the two is a vaccine?

9. Once blood enters the kidney through the renal artery, it delivers oxygen and nutrients to the kidney's cells and picks up their waste products. The following steps then happen. Put them into the proper order:

   a. Chemicals and water are absorbed back into the blood.
   b. The blood is filtered.
   c. Water and chemicals go into the renal pelvis, through the ureter, and into the bladder.
   d. Nutrients, water, and chemicals travel through a nephron.

For Questions 10 through 14, match the glands with their function.

10. hypothalamus
11. thyroid
12. pituitary
13. adrenal
14. pancreas

a. Produces cortisol, which causes the liver to release glucose into the blood
b. Controls the pituitary gland
c. Produces insulin, which enables glucose to enter the cells
d. Produces hormones that control many of the endocrine glands
e. Produces hormones that regulate the basal metabolic rate

15. If many of the endocrine glands in the body begin to malfunction, just one gland might be responsible. Which gland might that be?

TEST FOR MODULE #16

1. Define the following terms:

a. Autonomic nervous system
b. Sensory nervous system
c. Motor nervous system

For Questions 2 through 18, match the structure on the left with its appropriate description on the right

2. neurons	a. carry signals towards a neuron's cell body
3. neuroglia	b. composed of all nerves running off of the spinal cord
4. dendrites	c. composed of cell bodies, dendrites, and axons
5. axons	d. allows the two hemispheres of the brain to communicate
6. synapse	e. sensitive to salty, bitter, sweet, sour, and umami.
7. neurotransmitters	f. converts the rocking motion of the ossicles into electrical signals that the brain interprets as sound
8. central nervous system	g. support the neurons by performing various tasks so the neurons can do their job
9. peripheral nervous system	h. controls most high-level thinking skills such as reasoning
10. gray matter	i. the part of the autonomic nervous system that speeds up the heart rate
11. corpus callosum	j. deforms the lens in the eye to adjust focus
12. cerebellum	k. chemicals that travel across the synapse, transmitting a signal from the end of an axon to a receiving cell
13. cerebrum	l. a gap between the axon of a neuron and the receiving cell
14. sympathetic division	m. the part of the autonomic nervous system that slows the heart rate
15. parasympathetic division	n. controls the movement of voluntary muscles
16. taste buds	o. composed of the brain and the spinal cord
17. cochlea	p. composed mostly of neuron cell bodies
18. ciliary muscle	q. carry signals away from a neuron's cell body

SOLUTIONS TO THE TEST FOR MODULE #1

1. a. (1 pt) <u>Science</u> – An endeavor dedicated to the accumulation and classification of observable facts in order to formulate general laws about the natural world

b. (1 pt) <u>Papyrus</u> – An ancient form of paper, made from a plant of the same name

c. (1 pt) <u>Spontaneous generation</u> – The idea that living organisms can be spontaneously formed from non-living substances

2. (1 pt) <u>k</u>

3. (1 pt) <u>e</u>

4. (1 pt) <u>p</u>

5. (1 pt) <u>n</u>

6. (1 pt) <u>j</u>

7. (1 pt) <u>d</u>

8. (1 pt) <u>a</u>

9. (1 pt) <u>f</u>

10. (1 pt) <u>i</u>

11. (1 pt) <u>m</u>

12. (1 pt) <u>l</u>

13. (1 pt) <u>o</u>

14. (1 pt) <u>h</u>

15. (1 pt) <u>b</u>

16. (1 pt) <u>g</u>

17. (1 pt) <u>c</u>

18. (1 pt) <u>The progress of science depends on government and culture.</u>

19. (1 pt) <u>A Christian worldview</u> caused science to progress at the end of the Dark Ages.

20. (1 pt) <u>We should believe scientific ideas because of *evidence*, not because of the people who believe in them</u>.

Total points: 22

SOLUTIONS TO THE TEST FOR MODULE #2

1. a. (1 pt) Counter-example – An example that contradicts a conclusion

    b. (1 pt) Hypothesis – An educated guess that attempts to explain an observation or answer a question

    c. (1 pt) Theory – A hypothesis that has been tested with a significant amount of data

    d. (1 pt) Scientific law – A theory that has been tested by and is consistent with generations of data

2. (6 points – one for each in the correct order.)

    c. Make observations
    e. Form a hypothesis
    d. Perform experiments to confirm the hypothesis
    b. Hypothesis is now a theory
    f. Perform many experiments over several years
    a. Theory is now a law

3. (1 pt) The statement does not mention air. The proper form of the statement would be:

    In the absence of air, all objects, regardless of their weight, fall at the same rate.

4. (1 pt) Yes, he did. He made an observation, formed a hypothesis, and performed several experiments to test his hypothesis. It doesn't matter that his conclusion was wrong. He followed the scientific method.

5. (3 pts – one for the observation, one for the hypothesis, and one for the experiment) He observed that a penny sank in water. This led him to the hypothesis that all solid objects sink in water. He performed experiments in which he threw many objects into the water, all of which sank. Thus, the hypothesis was confirmed.

6. (1 pt) The other student provided a counter-example to show that the theory was wrong.

7. (1 pt) This story is similar because the eccentric student's theory as well as the theory that there were canals on Mars were both produced by the scientific method but were both wrong.

8. (1 pt) Scientists were quite surprised because a scientific theory said that they should not exist.

9. (3 pts – one for each limitation)
    a. It cannot prove anything.
    b. It is not 100% reliable
    c. It must conform to the scientific method

10. (1 pt) Yes, science can be used to study any question, as long as the scientific method is used.

Total points: 22

INSTRUCTIONS TO THE PARENT/HELPER IN EXPERIMENT 2.3

You need to make the flashlight stop working in some way. You will find several ideas below, as well as tips you can give the student if the student is stuck. Please give the tips to the student *only* if he is stuck. Once you have stopped the flashlight from working, give it to the student so that he can start the experiment.

1. (This one is relatively easy.) Put a piece of paper between the two batteries, so that the batteries no longer make electrical contact with one another. The student will probably figure this one out as soon as he or she takes the flashlight apart. The experiment that the student designs to check the hypothesis should involve removing the paper to see that the flashlight is working, and then putting it back in to see that it stops working again. This will ensure that it was the paper causing the problem.

2. (This one is a little harder.) Put a *single* dead battery in the flashlight. Since your flashlight probably uses at least 2 batteries, that means there will be at least one good battery in the flashlight. It will probably be easy for the student to form the hypothesis, but it will be much harder to test. The student's experiment to test the hypothesis would have to involve replacing the batteries one at a time to determine *which* is the dead one. Simply replacing all of the batteries and getting the flashlight working is NOT ENOUGH. The student must determine WHICH battery was dead.

3. (This one is a little harder than the first two.) If you have a flashlight with a bad bulb, use that flashlight. If you want to break the bulb (you can do this by dropping the flashlight a few times) in the flashlight you are using, that would work, too. In order for the student to be successful, however, you must have access to a good bulb. The experiment that the student would do should involve putting in a new bulb to fix the flashlight and then putting the old bulb back in to see that the flashlight stops working again.

4. (This is the hardest one.) If you unscrew the top of the flashlight and look at where the top touches the battery, you will see a little metal clip. You can use masking tape to cover that clip and then use a marker to color it the same color as the surrounding material. This will camouflage what you have done. When the student discovers the problem, the experiment to confirm that it is the problem should involve removing it and seeing that the flashlight works followed by putting it back on to see that the flashlight no longer works.

Of course, if you can think of other ways to sabotage the flashlight, please use them. Remember, however, that **it is not enough for the student to just get the flashlight working**. It is commonplace to take something that was broken apart, put it back together, and find that it is suddenly working again. Thus, if the student just fixes the flashlight, he or she will not know for sure that the hypothesis was correct. Something else might have happened to accidentally fix the flashlight. That's why the student must perform an experiment that *demonstrates* he correctly identified the problem.

SOLUTIONS TO THE TEST FOR MODULE #3

1. a. (1 pt) <u>Experimental variable</u> – An aspect of an experiment that changes during the course of the experiment

b. (1 pt) <u>Control (of an experiment)</u> – The variable or part of the experiment to which all others will be compared

c. (1 pt) <u>Blind experiments</u> – Experiments in which the participants do not know whether or not they are a part of the control group

d. (1 pt) <u>Double-blind experiments</u> – Experiments in which neither the participants nor the people analyzing the results know who is in the control group

2. (1 pt) <u>Experimental variables can affect the results of your experiment</u>. The experimental variable from which you can learn something should be kept, and the rest should be reduced or eliminated so that they do not throw off the results of the experiment.

3. (3 pts – one for each correct answer. Subtract one for every wrong letter) <u>Items (b), (c), and (e) are</u> <u>experimental variables</u>. They used the same dishwasher and detergent throughout, so neither of those is variable. The problem specifically states that they used different glasses and different additives. Since they used the same dishwasher again and again, it could very well be dirty from a previous load.

4. (1 pt) <u>You will learn something from (b)</u>.

5. (2 pts – one for each correct answer. Subtract one for every wrong letter) <u>Items (c) and (e) should</u> <u>be reduced or eliminated</u>.

6. (1 pt) <u>The data are subjective</u>. It depends on the opinions of those inspecting the glasses, since there is no number that you can put on the level of cleanliness of dishes.

7. (2 pts – one for the right thing that will happen to each) <u>When soap is put in the water, the paper</u> <u>clip will sink, but the cork will not</u>. The paper clip floats because of surface tension. You know that because, since it is made of metal, it is denser than water. The only way it could float, then, is by surface tension. As you learned in Experiment 3.4, soap reduces surface tension, and it would cause the paper clip to sink. The cork floats not because of surface tension, but because it is less dense than water. Thus, it will keep floating regardless of surface tension.

8. (1 pt) <u>This should be a double-blind experiment</u>. If the participants know that they are getting a supposed energy boost, it might make them feel better mentally. This could make them feel less tired, regardless of whether the stuff in the bar did anything for them. The investigator should be blind as well, because listening to what people say and interpreting it is subjective.

9. (1 pt) <u>The radiation level without any lead is about 180 counts per second</u>. No lead would be the case when the dot is at 0 on the x-axis. That dot is on the far left, and it is about at a y-axis value of 180. The student's answer can be as low as 175.

10. (1 pt) 50 counts per minute is halfway between 40 and 60 on the y-axis. The dot that has that vertical position is essentially at <u>2 millimeters</u> on the x-axis. You could have said anything between 1½ and 2.

11. (1 pt) After a thickness of 4 millimeters, no matter how far to the right you go, the dots don't go down much further at all. Thus, you don't gain much more protection after <u>4 millimeters</u>, so you might as well not waste any more money. You could have said 5 or 6 as well.

Total points: 18

SOLUTIONS TO THE TEST FOR MODULE #4

1. a. (1 pt) Simple machine – A device that either multiplies or redirects a force

b. (1 pt) Force – A push or pull exerted on an object in an effort to change that object's velocity

c. (1 pt) Mechanical advantage - The amount by which force or motion is magnified in a simple machine

d. (1 pt) Diameter – The length of a straight line that travels from one side of a circle to another and passes through the center of the circle

e. (1 pt) Circumference – The distance around a circle, equal to 3.1416 times the circle's diameter

2. (2 pts – one for each correct answer. Subtract one for every wrong letter.) Experiments (a) and (b) are applied science experiments, because the goal is specifically to make something better.

3. (2 pts – one for each correct answer. Subtract one for every wrong letter.) Items (c) and (d) are technology, because they make life better in some way.

4. (1 pt) The mechanical advantage of a lever increases the closer the fulcrum is to the resistance. Thus, move the fulcrum closer to the rock.

5. (1 pt) You apply a force at the handle of the scissors. Thus, the effort is at one end. The resistance is the thing you are cutting. In between is the little rivet that does not move. That's the fulcrum. Thus, the fulcrum is between the resistance and the effort, so scissors are composed of two first-class levers.

6. (1 pt) It means that the speed at which the resistance moves will be magnified by 5.

7. (1 pt – give ½ if the student has the right equation but the wrong answer.) A wheel and axle's mechanical advantage is given by:

Mechanical advantage = (diameter of the wheel) ÷ (diameter of the axle)

Mechanical advantage = 20 ÷ 2 = 10

8. (2 pts – one for the force being magnified, and one for the drawback.) In a wheel and axle, turning the wheel magnifies the effort. Thus, the force with which the wheel is turned is magnified. The drawback is that you must turn the wheel more than the axle turns. The student can also say that the axle turns more slowly than the wheel.

9. (1 pt) The mechanical advantage of a multiple-pulley system (which is what a block and tackle is) is equal to the number of pulleys. Thus, the mechanical advantage is 2.

10. (1 pt) As diagrammed, this is a wedge. See Figure 4.7.

11. (1 pt – give ½ if the student has the right equation but the wrong answer.) Whether the machine is an inclined plane or a wedge, the mechanical advantage is the same:

$$\text{Mechanical advantage} = (\text{length of slope}) \div (\text{height})$$

$$\text{Mechanical advantage} = 10 \div 2 = \underline{5}$$

12. (2 pts – one for the circumference and one for the mechanical advantage. Even if the answers are wrong, give ½ for each correct equation.) To determine the mechanical advantage of a screw, we must first know the circumference of what's being grabbed. Since no screwdriver is mentioned, we must assume it is the head of the screw:

$$\text{Circumference} = 3.1416 \text{ x } (\text{diameter})$$

$$\text{Circumference} = 3.1416 \text{ x } 0.1 = 0.31416$$

Now we can use the mechanical advantage equation for a screw:

$$\text{Mechanical advantage} = (\text{circumference}) \div (\text{pitch})$$

$$\text{Mechanical advantage} = 0.31416 \div 0.05 = \underline{6.2832}$$

13. (2 pts – one for the circumference and one for the mechanical advantage. Even if the answers are wrong, give ½ for each correct equation.) The mechanical advantage is different now, because a screwdriver is being used. For that, we need to know the circumference of the screwdriver:

$$\text{Circumference} = 3.1416 \text{ x } (\text{diameter})$$

$$\text{Circumference} = 3.1416 \text{ x } 1 = 3.1416$$

Now we can use the mechanical advantage equation for a screw:

$$\text{Mechanical advantage} = (\text{circumference}) \div (\text{pitch})$$

$$\text{Mechanical advantage} = 3.1416 \div 0.05 = \underline{62.832}$$

Total points: 22

SOLUTIONS TO THE TEST FOR MODULE #5

1. a. (1 pt) Life science – A term that encompasses all scientific pursuits related to living organisms

b. (1 pt) Archaeology – The study of past human life as revealed by preserved relics

c. (1 pt) Geology – The study of earth's history as revealed in the rocks that make up the earth

d. (1 pt) Known age – The age of an artifact as determined by a date printed on it or a reference to the artifact in a work of history

e. (1 pt) Dendrochronology – The process of counting tree rings to determine the age of a tree

f. (1 pt) The Principle of Superposition – When artifacts are found in rock or earth that is layered, the deeper layers hold the older artifacts.

2. (1 pt) The bibliographic test

3. (1 pt) The internal test

4. (1 pt) The external test

5. (1 pt) One traces Mary's lineage while the other traces Joseph's.

6. (1 pt) The Bible passes the external and bibliographic tests better than any other document of its time.

7. (1 pt) Some passages may look like contradictions, but the apparent contradiction might be due to difficulties in translation. Thus, a contradiction must be ironclad in order to make an ancient document fail the internal test.

8. (1 pt) The archaeologist must find a master tree ring pattern in the log's rings. That way, he will know the age of that ring pattern and can count the rest of the rings to determine when the tree was cut down.

9. (1 pt) Known ages are more certain.

10. (1 pt) All dating methods (except the Principle of Superposition) give absolute ages.

11. (1 pt) The Principle of Superposition.

12. (1 pt) The Principle of Superposition assumes that in rock or soil that is layered, the layers were formed one at a time. This is not necessarily true.

Total points: 17

SOLUTIONS TO THE TEST FOR MODULE #6

1. a. (1 pt) <u>Catastrophism</u> – The view that most of earth's geological features are the result of large-scale catastrophes such as floods, volcanic eruptions, etc.

b. (1 pt) <u>Uniformitarianism</u> – The view that most of earth's geological features are the result of slow, gradual processes that have been at work for millions or even billions of years

c. (1 pt) <u>Humus</u> – The decayed remains of once-living creatures

d. (1 pt) <u>Minerals</u> – Inorganic crystalline substances found naturally in the earth

e. (1 pt) <u>Weathering</u> – The process by which rocks are broken down to form sediments

f. (1 pt) <u>Erosion</u> – The process by which rock and soil are broken down and transported away

g. (1 pt) <u>Unconformity</u> – A surface of erosion that separates one layer of rock from another

2. (1 pt) <u>Catastrophism</u> allows for a "young" earth, because in catastrophism, geological structures are formed rapidly.

3. (1 pt) The rock is <u>metamorphic</u>.

4. (1 pt) The rock is <u>sedimentary</u>.

5. (1 pt) The rock is <u>igneous</u>.

6. (1 pt) This is <u>chemical weathering</u>, because the iron changed to a new substance: rust.

7. (1 pt) <u>The first region will experience more erosion</u>, as plants tend to reduce the effect of erosion.

8. (1 pt) Without groundwater seeping in through the ceiling, <u>you will see neither stalactites nor stalagmites</u>. Remember, even stalagmites are formed from water that drips in from the ceiling.

9. A. (1 pt) <u>igneous rock</u>
    B. (1 pt) <u>sedimentary rock</u>
    C. (1 pt) <u>metamorphic rock</u>

10. (1 pt) Number <u>2</u> points to the angular unconformity.

11. (1 pt) Number <u>1</u> points to a disconformity.

12. (1 pt) Number <u>3</u> points to a nonconformity.

Total points: 20

SOLUTIONS TO THE TEST FOR MODULE #7

1. a. (1 pt) <u>Fossil</u> – The preserved remains of a once-living organism

   b. (1 pt) <u>Petrifaction</u> – The conversion of organic material into rock

   d. (1 pt) <u>Resin</u> – A thick, slowly flowing liquid produced by plants that can harden into a solid

   d. (1 pt) <u>Extinct</u> – A term applied to a species that was once living but now is not

2. (2 pts – one for casts, one for molds) <u>Casts and molds</u>

3. (2 pts – one for amber, one for ice) Typically, <u>creatures trapped in amber and creatures trapped in</u> <u>ice</u> provide the most detail, as they usually have many of the soft parts as well as the hard parts of the creature preserved.

4. (1 pt) <u>Casts</u> depend on the fact that a mold formed previously.

5. (1 pt) <u>Petrified remains</u>

6. (1 pt) Plants usually form <u>carbonized remains</u> because their tissues are well-suited for it.

7. (4 pts – one for each feature) The four general features of the fossil record are:

   I. <u>Fossils are usually found in sedimentary rock. Since most sedimentary rock is laid down</u> <u>by water, it follows that most fossils were laid down by water.</u>
   II. <u>The vast majority of the fossil record is made up of hard-shelled creatures like clams. Most</u> <u>of the remaining fossils are of either water-dwelling creatures or insects. Only a tiny, tiny</u> <u>fraction of the fossils we find are of plants, reptiles, birds, and mammals.</u>
   III. <u>Many of the fossils we find are of organisms that are still alive today. Many of the fossils</u> <u>we find are of organisms that are now extinct.</u>
   IV. <u>The fossils found in one layer of stratified rock can be considerably different from the</u> <u>fossils found in another layer of stratified rock.</u>

8. (1 pt) <u>The geologist is a uniformitarian.</u> Although many catastrophists do believe that the lowest layers of rock represent the Creation Week and the time before the Flood and that the upper layers represent mostly the Flood, uniformitarians believe each and every layer was laid down during a different period of earth's history.

9. (1 pt) <u>The geologist is a catastrophist.</u> Most catastrophists believe that most fossils were laid down during the Flood.

10. (1 pt) <u>No</u>, you cannot believe the earth is only a few thousand years old if you are a uniformitarian. Uniformitarians need hundreds of millions of years for the sedimentary rocks we see today to be formed by the processes we see occurring today.

Total points: 18

SOLUTIONS TO THE TEST FOR MODULE #8

1. a. (1 pt) <u>Index fossils</u> – Fossils that are assumed to represent a certain period in earth's past

b. (1 pt) <u>Geological column</u> – A theoretical picture in which layers of rock from around the world are meshed together into a single, unbroken record of earth's past

c. (1 pt) <u>The Theory of Evolution</u> – A theory stating that all life on this earth has one (or a few) common ancestor(s) that existed a long time ago

2. (1 pt) <u>Layer B</u> and layer F represent the same time period, because they have the same index fossils.

3. (1 pt) <u>Layer C</u> and layer G represent the same time period, because they have the same index fossils.

4. (1 pt) <u>Layer E</u> and layer H represent the same time period, because they have the same index fossils.

5. (1 pt) Layer F is at the bottom of formation 2, so it is the oldest layer in that formation. Layer A is at the bottom of formation 1, so it is the oldest layer in that formation. However, the question asks which layer is the very oldest. Well, since layer F and layer B have the same index fossils, it means they represent the same time period. Since layer A is under layer B, it means it is older than layer B. This means A is older than F, because F represents the same time period as B. Thus, <u>A is the oldest layer</u>.

6. (2 pts – one for each letter. Subtract one for any wrong letters.) <u>The time periods represented by layers A and D are not in formation 2</u>, because the index fossils for those time periods are not in formation 2.

7. (1 pt) Since the squares appear lower in the formations than the smiley faces, uniformitarians would assume that <u>the creatures whose fossils are represented by squares came first</u>.

8. (1 pt) <u>The data from Mt. St. Helens indicate that the major geological formations we see today can be formed quickly as a result of catastrophes.</u>

9. (1 pt) <u>No</u>. The geological column is a theoretical construct.

10. (1 pt) <u>It supports the idea that God created each kind of plant and animal individually.</u> It indicates that plants and animals are unique and there are no intermediate links between them. This is evidence that the plants and animals were created individually.

11. (2 pts – one for what it is and one for the fact that uniformitarians must use it) <u>A paraconformity is an unconformity for which no evidence exists. Some geologists nevertheless assume that it does. Uniformitarians require their existence</u> to deal with situations where index fossils from different time periods appear in a single layer of rock.

12. (2 pts – one for each problem) The text discusses at least six problems. The student need only list two.

a. <u>There are too many fossils in the fossil record.</u>

b. <u>Fossils such as the *Tyrannosaurus rex* bone that contains soft tissue are hard to understand in the uniformitarian framework.</u>

c. <u>Fossil graveyards with fossils from many different climates are hard to understand in the uniformitarian view.</u>

d. <u>Index fossils are called into question by the many creatures we once thought were extinct but we now know are not.</u>

e. <u>Uniformitarians must assume the existence of paraconformities.</u>

f. <u>Uniformitarians must believe that evolution occurred, and there is no evidence for evolution.</u>

13. (2 pts – one for each problem) The text discusses at least three problems. The student need only list 2.

a. <u>Catastrophists have offered no good explanation for the existence of unconformities between rock layers laid down by the Flood.</u>

b. <u>Catastrophists cannot explain certain fossil structures that look like they were formed under "normal" living conditions which would not exist during the Flood.</u>

c. <u>Catastrophists have not yet explained the enormous chalk deposits we find in terms of the Flood.</u>

14. (1 pt) <u>No</u>. Fossilized hats, legs in boots, and waterwheels tell us that fossils can form rapidly.

Total points: 20

SOLUTIONS TO THE TEST FOR MODULE #9

1. a. (1 pt) <u>Atom</u> – The smallest chemical unit of matter

b. (1 pt) <u>Molecule</u> – Two or more atoms linked together to make a substance with unique properties

c. (1 pt) <u>Photosynthesis</u> – The process by which green plants and some other organisms use the energy of sunlight and simple chemicals to produce their own food

d. (1 pt) <u>Metabolism</u> – The sum total of all processes in an organism that convert energy and matter from outside sources and use that energy and matter to sustain the organism's life functions

e. (1 pt) <u>Receptors</u> – Special structures that allow living organisms to sense the conditions of their internal or external environment

f. (1 pt) <u>Cell</u> – The smallest unit of life in creation

2. (2 pts – one for each criterion) The two missing criteria are:

- <u>All life forms have a method by which they extract energy from the surroundings and convert it into energy that sustains them.</u>

- <u>All life forms reproduce.</u>

3. (1 pt – ½ for each arrow) <u>Arrow (a) is pointing to a nucleotide, and arrow (b) is pointing to the backbone.</u>

4. (1 pt) <u>Information is stored in DNA as a sequence of nucleotide bases.</u> Just like this entire book could be sent as a sequence of dots and dashes in Morse code, all of the information of life can be stored as a sequence of nucleotide bases.

5. (3 pts – ½ for each) Since only adenine and thymine link up, and since only cytosine and guanine link up, the other half of the DNA must be:

<u>adenine, guanine, cytosine, thymine, cytosine, adenine</u>

6. (2 pts – one for each) Photosynthesis produces <u>glucose and oxygen</u>.

7. (1 pt) <u>water</u>

8. (1 pt) <u>Organisms that produce lots of offspring tend to live very dangerous lives.</u> They need lots of offspring to "replace" those that die before having offspring.

9. (1 pt) <u>That person's cells are reproducing all of the time.</u> Thus, the person has reproduced countless times on the cellular level.

10. a. (1 pt) <u>membrane</u>

b. (1 pt) <u>nucleus</u>

c. (1 pt) <u>organelles</u>

Total points: 21

SOLUTIONS TO THE TEST FOR MODULE #10

1. a. (1 pt) <u>Prokaryotic cell</u> – A cell that has no distinct, membrane-bounded organelles

b. (1 pt) <u>Eukaryotic cell</u> – A cell with distinct, membrane-bounded organelles

c. (1 pt) <u>Pathogen</u> – An organism that causes disease

d. (1 pt) <u>Decomposers</u> – Organisms that break down the dead remains of other organisms

e. (1 pt) <u>Vegetative reproduction</u> – The process by which one part of a plant can form new roots and develop into a complete plant

2. (1 pt) <u>Protista</u>

3. (1 pt) Since it is made of eukaryotic cells, it is not in kingdom Monera. Since it is neither one-celled nor does it make its own food, it is not in kingdom Protista nor is it in kingdom Plantae. It eats living organisms, so it is not in kingdom Fungi. It must be in kingdom <u>Animalia</u>.

4. (1 pt) What it eats is irrelevant. Since it is a single prokaryotic cell, it belongs in kingdom <u>Monera</u>.

5. (1 pt) Organisms that make their own food and have specialized structures like roots, stems, and leaves belong in kingdom <u>Plantae</u>.

6. (1 pt) Since it is a single eukaryotic cell, you might think it belongs in kingdom Protista. However, it is a decomposer because it eats only dead organisms. Thus, it is a single-celled member of kingdom <u>Fungi</u>.

7. Salt will reduce bacterial growth and reproduction, and covering will reduce the new bacteria introduced onto the bacon.

      a. (1 pt) The <u>first</u> will have the most bacteria.

      b. (1 pt) The <u>third</u> will have the least.

8. (3 pts – one for the difference, and one for each kingdom) <u>Algae have no specialized structures, while plants do. Algae belong in kingdom Protista, while plants belong in kingdom Plantae.</u>

9. (1 pt) <u>Yes</u>, decomposers recycle dead matter so that it can be used by living organisms again.

10. (1 pt) <u>The man did not really get rid of the mushroom fungus. The mycelium is still underground, so mushrooms will continually be produced.</u>

11. (1 pt) <u>No</u>, a wilting plant is not necessarily dead. It has just lost turgor pressure. Add water, and the wilting may go away. Plants will wilt before they die, but just being wilted does not necessarily mean the plant has died. Many wilted plants live for a long time if they get water soon enough.

12. (2 pts – one for each) The <u>central vacuole</u> and the <u>cell wall</u> work together to make turgor pressure. The central vacuole expands by filling with water, pushing the contents of the cell against the cell wall. The cell wall pushes back, making the pressure.

Total points: 20

SOLUTIONS TO THE TEST FOR MODULE #11

1. a. (1 pt) <u>Axial skeleton</u> – The portion of the skeleton that supports and protects the head, neck, and trunk

b. (1 pt) <u>Appendicular skeleton</u> – The portion of the skeleton that attaches to the axial skeleton and has the limbs attached to it

c. (1 pt) <u>Exoskeleton</u> – A body covering, typically made of chitin, that provides support and protection

d. (1 pt) <u>Symbiosis</u> – A close relationship between two or more species where at least one benefits

2. (1 pt) <u>e</u>

3. (1 pt) <u>i</u>

4. (1 pt) <u>b</u>

5. (1 pt) <u>j</u>

6. (1 pt) <u>a</u>

7. (1 pt) <u>c</u>

8. (1 pt) <u>h</u>

9. (1 pt) <u>f</u>

10. (1 pt) <u>d</u>

11. (1 pt) <u>g</u>

12. (1 pt) The larger the range of motion, the less the stability of a joint. Thus, the <u>second joint is probably more stable.</u>

13. (1 pt) Phototropism is the tendency of plants to grow towards the light. That was exhibited by the <u>second plant</u>. The first plant simply demonstrates that plants need light to survive.

14. (1 pt) <u>This person is more likely to get sick because sweat helps to feed bacteria and fungi that fight off pathogenic organisms. Without the sweat, these beneficial organisms would die, and the person will be more likely to be infected by pathogenic organisms.</u>

15. a. (1 pt) If an animal has feathers, it is a <u>bird</u>. It is irrelevant whether or not the creature can fly. The presence of feathers means it's a bird.

b. (1 pt) If an animal has hair, it is a <u>mammal</u>.

c. (1 pt) If it breathes through its skin, it is most likely an <u>amphibian</u>.

Total points: 20

SOLUTIONS TO THE TEST FOR MODULE #12

1. a. (1 pt) <u>Producers</u> – Organisms that produce their own food

b. (1 pt) <u>Consumers</u> – Organisms that eat living producers and/or other consumers for food

c. (1 pt) <u>Herbivore</u> – A consumer that eats producers exclusively

d. (1 pt) <u>Carnivore</u> – A consumer that eats only other consumers

e. (1 pt) <u>Omnivore</u> – A consumer that eats both producers and other consumers

f. (1 pt) <u>Basal metabolic rate</u> – The minimum amount of energy required by the body in a day

2. a. (1 pt) Yeast is in kingdom Fungi, so it is a <u>decomposer</u>.

b. (1 pt) An ant cannot make its own food, so it is a <u>consumer</u>.

c. (1 pt) A fly cannot make its own food, so it is a <u>consumer</u>.

d. (1 pt) Corn is a plant, so it is a <u>producer</u>.

e. (1 pt) A rose bush is a plant, so it is a <u>producer</u>.

3. (4 pts – one for each) Combustion requires <u>oxygen</u> and produces <u>carbon dioxide</u>, <u>water</u>, and <u>energy</u>.

4. (1 pt) <u>e</u>

5. (1 pt) <u>g</u>

6. (1 pt) <u>f</u>

7. (1 pt) <u>a</u>

8. (1 pt) <u>d</u>

9. (1 pt) <u>b</u>

10. (1 pt) <u>c</u>

11. (1 pt) <u>Endothermic organisms must expend a lot of energy keeping their body temperature constant</u>. Since this is independent of activity, it is a part of the BMR.

12. (1 pt) Without the proper proteins, <u>a person's cells cannot manufacture the proteins they need to manufacture</u>. In order to make proteins, cells need amino acids. There are 8 amino acids our cells

cannot make. Thus, we must eat them. If we don't, our cells run out and cannot make proteins that have those amino acids in them.

13. (1 pt) <u>Saturated fats</u> are solid at room temperature.

Total points: 25

SOLUTIONS TO THE TEST FOR MODULE #13

1. a. (1 pt) <u>Digestion</u> – The process by which an organism breaks down its food into small units that can be absorbed by the body

b. (1 pt) <u>Vitamin</u> – A chemical substance the body needs in small amounts to stay healthy

2. a. (1 pt) <u>salivary glands</u>

b. (1 pt) <u>esophagus</u>

c. (1 pt) <u>stomach</u>

d. (1 pt) <u>large intestine</u>

e. (1 pt) <u>small intestine</u>

f. (1 pt) <u>rectum</u>

g. (1 pt) <u>liver</u>

3. (1 pt) <u>The small intestine</u> is where most nutrient absorption occurs.

4. (1 pt) <u>The stomach</u> turns the bolus into chyme.

5. (1 pt) <u>The large intestine</u> converts waste into feces.

6. (1 pt) <u>The pancreas</u> produces sodium bicarbonate.

7. (1 pt) <u>The liver</u> produces bile.

8. (1 pt) <u>The salivary glands</u> produce saliva.

9. (1 pt) <u>The tongue</u> moves the food in the mouth to form the bolus.

10. (1 pt) <u>The epiglottis covers the larynx when you swallow to make sure that food goes down the esophagus only.</u>

11. a. (1 pt – ½ for each. Count ½ off for every wrong one listed.) <u>Vitamins C and the B vitamins</u>. A, D, E and K are the fat-soluble vitamins. If a vitamin is not fat-soluble, it is water-soluble.

b. (1 pt – ½ for each. Count ½ off for every wrong one listed.) <u>Vitamins D and K</u>. Vitamin D can be made by the body through sunlight hitting the skin, and vitamin K is made by bacteria in the large intestine.

c. (1 pt – one-third for each. Count one-third off for every wrong one listed.) <u>Vitamins D, E, and K</u>. The fat-soluble vitamins are the ones that build up easily in the body. These are the fat-soluble ones on the list.

Total points: 20

SOLUTIONS TO THE TEST FOR MODULE #14

1. a. (1 pt) <u>Veins</u> – Blood vessels that carry blood back to the heart

b. (1 pt) <u>Arteries</u> – Blood vessels that carry blood away from the heart

c. (1 pt) <u>Capillaries</u> – Tiny, thin-walled blood vessels that allow the exchange of gases and nutrients between the blood and cells and are located between arteries and veins

2. (3 pts – 1 for the proper order and ½ for listing each chamber's name) Blood flows into the <u>right atrium</u> and then gets dumped into the <u>right ventricle</u>. It then goes to the lungs and returns oxygenated. The oxygenated blood is dumped into the <u>left atrium</u> and then gets dumped into the <u>left ventricle</u> before being pumped to the body.

3. (1 pt) Deoxygenated blood leaves the <u>right ventricle</u> on its way to the lungs.

4. (2 pts – one for artery, one for the fact that you can't be certain) Most likely, the blood came from an <u>artery</u>, because most arteries contain oxygenated blood. <u>You cannot be 100% certain</u>, however, because there are exceptions.

5. (1 pt) <u>Plasma</u> makes up the majority of the blood.

6. (2 pts – one for what it is, one for the fact that it is aided by platelets) <u>Blood clotting is the process by which blood seals wounds to keep it from leaking out an injured artery. It is aided by the blood platelets</u>.

7. (1 pt) <u>Red blood cells</u> contain hemoglobin.

8. (1 pt) <u>White blood cells</u> fight disease.

9. (1 pt) You are looking at <u>alveoli</u>.

10. (1 pt) Exhaled air travels from the lungs, into the trachea, up the larynx, to the pharynx, and out the mouth or nasal cavity. Thus, it passes through the <u>trachea</u> first.

11. (1 pt) Volume is controlled by how much air passes over the vocal cords. Thus, <u>the man</u> is passing more air over his vocal cords than is the woman.

12. (1 pt) You are looking at <u>capillaries</u>. Those are the only vessels in which oxygen exchange occurs.

13. (1 pt) <u>Xylem are tubes in plants that transport water up the plant</u>.

14. (1 pt) Blood cells are produced in <u>bone marrow</u>.

Total Points: 20

SOLUTIONS TO THE TEST FOR MODULE #15

1. a. (1 pt) Gland – A group of cells that prepare and release a chemical for use by the body

b. (1 pt) Vaccine – A weakened or inactive version of a pathogen that stimulates the body's production of antibodies that can destroy the pathogen

c. (1 pt) Hormone – A chemical messenger released into the bloodstream that sends signals to distant cells, causing them to change their behavior in specific ways

2. (1 pt) The urinary system

3. (1 pt) The lymphatic system

4. (1 pt) The lymphocytes are found in the lymph nodes.

5. (1 pt) Lymph is pumped through lymph vessels by the contraction of muscles in the body.

6. (1 pt) B-cells make antibodies.

7. (1 pt) Memory B-cells give the lymphatic system a "memory" of infections. Vaccines use this memory feature to give the body immunity to diseases it hasn't actually fought yet.

8. (1 pt) The second is the vaccine. Vaccines are used to give a person immunity to a disease he has not had yet. Only in very rare cases can a vaccine be used as a treatment for a disease.

9. (4 pts – take one off for each step that is out of order) The order is b, d, a, c.

10. (1 pt) b

11. (1 pt) e

12. (1 pt) d

13. (1 pt) a

14. (1 pt) c

15. (1 pt) The pituitary gland might be malfunctioning. Since it controls most endocrine glands, a failure there would make most endocrine glands fail. If the student says hypothalamus, count it correct, since the hypothalamus controls the pituitary gland.

Total points: 20

SOLUTIONS TO THE TEST FOR MODULE #16

1. a. (1 pt) <u>Autonomic nervous system</u> – The system of nerves that carries instructions from the CNS to the body's smooth muscles, cardiac muscle, and glands

b. (1 pt) <u>Sensory nervous system</u> – The system of nerves that carries information from the body's receptors to the CNS

c. (1 pt) <u>Motor nervous system</u> – The system of nerves that carries instructions from the CNS to the skeletal muscles

2. (1 pt) <u>c</u>

3. (1 pt) <u>g</u>

4. (1 pt) <u>a</u>

5. (1 pt) <u>q</u>

6. (1 pt) <u>l</u>

7. (1 pt) <u>k</u>

8. (1 pt) <u>o</u>

9. (1 pt) <u>b</u>

10. (1 pt) <u>p</u>

11. (1 pt) <u>d</u>

12. (1 pt) <u>n</u>

13. (1 pt) <u>h</u>

14. (1 pt) <u>i</u>

15. (1 pt) <u>m</u>

16. (1 pt) <u>e</u>

17. (1 pt) <u>f</u>

18. (1 pt) <u>j</u>

Total points: 20

QUARTERLY TEST #1

1. Define the following terms:

a. Spontaneous generation
b. Counter example
c. Hypothesis
d. Control (of an experiment)
e. Double-blind experiments
f. Simple machine
g. Mechanical advantage

2. What was Gregor Mendel's greatest contribution to science?

3. Max Planck was the first to come up with an assumption that ushered in a revolution in modern science. What was that assumption?

4. Copernicus wrote a book that advocated the system that is considered the best way of understanding the sun and planets. What is that view called? Don't use his name as a part of your answer.

5. What do we call the people who want to turn lead (or other inexpensive items) into gold (or other valuable items)?

6. Which scientist laid down the laws of motion, developed a universal law of gravity, invented calculus, wrote many commentaries on the Bible, showed white light is really composed of many different colors of light, and came up with a completely different design for telescopes?

7. Dr. Francis S. Collins is one of the today's greatest geneticists. His studies have led him to believe that the genetic code is the result of God's design. Should you believe this because Dr. Collins says so?

8. A scientist makes some observations and forms an idea that explains those observations. He then does some experiments to test his idea, and all his experiments support the idea. At this point, is the idea a hypothesis, a theory, or a scientific law?

9. A bowling ball and piece of paper are anchored to the top of a tall chamber that has no air in it. At the same instant, they are both released so they start to fall. Which hits the bottom of the chamber first?

10. Can science be used to prove that a given idea is correct?

11. A scientist finds that he must either discard his hypothesis or modify it. What has happened to make him come to this conclusion?

12. What theory is Percival Lowell best remembered for?

13. Can you use science in an attempt to determine whether or not ghosts exist?

Questions 14 through 16 refer to the following story:

A scientist is testing the effectiveness of a pill that claims it will prevent people from catching colds. The scientist splits her volunteers into two groups: A and B. She gives pills made of sugar to group A, and she gives the pill she is testing to group B. She has all the volunteers keep a daily diary, noting how they felt each day and whether or not they were experiencing cold-like symptoms. After six months, she gathers the diaries and reads through them.

14. Which group was the control?

15. Should this be a single-blind experiment, a double-blind experiment, or neither?

16. Suppose the volunteers in group B experienced significantly fewer cold symptoms than the volunteers in group A. What would that tell you about he drugs effectiveness?

Questions 17 through 19 refer to the following story:

A weatherman decides to see how the temperature varies over the area in which he lives. He sets up eleven thermometers at various distances from the downtown office where he works. He then records the temperature every day for a month and graphs the results as shown on the right.

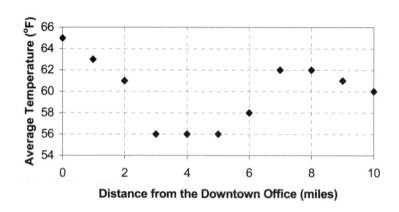

17. What can you say about the temperature at the weatherman's office compared to all the others?

18. What was the average temperature 6 miles from the downtown office?

19. If you wanted to stay in the coolest part of the weatherman's area, how far from his office should you be?

20. In a first-class lever, the fulcrum is 10 inches from the resistance and 100 inches from the effort. What is the mechanical advantage?

21. In the situation described in #20, what does the mechanical advantage do?

22. Suppose you use a block and tackle system with four pulleys that work together. What is the mechanical advantage?

23. If you want to lift a load 10 feet into the air using the block and tackle system in #22, how many feet of rope will you have to pull?

24. If you are having trouble getting a screw into a board, should you use a longer screwdriver or a fatter one?

25. What were the ancient Egyptians known for: science, applied science, or technology?

QUARTERLY TEST #2

1. Define the following terms:

a. Aristotle's dictum
b. Dendrochronology
c. Minerals
d. Unconformity
e. Petrifaction
f. Resin
g. Index fossils
h. Geological column

2. A historian is trying to determine whether or not an ancient manuscript is consistent with the archaeological records of its time. Is the historian using the internal, external, or bibliographic test?

3. A historian is looking at two different books of ancient history. We do not have the original of either one. The earliest copy of the first book was made 500 years after the original was written, and there are 22 other independent copies that are essentially identical to it. The earliest copy of the second book was made 1,000 years after the original was written, and there are 10 other copies that are essentially identical to it. Which of these two books is most likely to be true to the original work?

4. An archaeologist find a coin with the date it was made stamped on it. Does this coin have a known age or an absolute age?

5. The same archaeologist is working in a layer of soil above the layer in which he found the coin. He finds some pottery, but there are no dates on it, and there is no dating method he can use on it. However, he still decides he knows something about the age of the pottery. What does he decide he knows about the age of the pottery?

6. What principle did the archaeologist in #5 use?

7. What historical evidence (besides the fact that it is mentioned in the Bible) do we have for the fact that a global flood took place sometime in earth's past?

8. Which view of geology (uniformitarianism or catastrophism) can accommodate both a young earth and an old earth?

9. If a rock formed from solidified magma, what kind of rock is it?

10. A limestone rock gets hit with acid rain and bubbles, producing a gas. Does the rock experience chemical or physical weathering?

11. Suppose you find an unconformity between two layers of sedimentary rock. If the layers are parallel to one another, which kind of unconformity is it?

12. Suppose a geologist concludes that an unconformity *must* exist in a layer of rock, despite the fact that there is no evidence for one. What kind of unconformity is it?

13. A river flows through a dense forest where its banks are lined with lush greenery. Eventually, it makes its way to an area that has been cleared of all plants so that construction can begin. In which area would you expect less erosion?

14. In what kind of rock (igneous, metamorphic, or sedimentary) do you typically find fossils?

15. If scientists tell you that an animal is extinct, can you be sure they are right?

16. If an animal falls into a lake full of mineral-rich water, what is the most likely fossil that will form (assuming one forms at all)?

17. You split open a rock to find what appears to be an incredibly detailed drawing of a leaf. What kind of fossil have you found?

18. According to uniformitarians, what determined the type of fossils you find in a layer of rock?

19. According to catastrophists, what determined the type of fossils you find in a layer of rock?

Questions 20 through 22 refer to the diagram on the right, where two different geological formations are shown. The symbols represent index fossils, and the letters are just for identification.

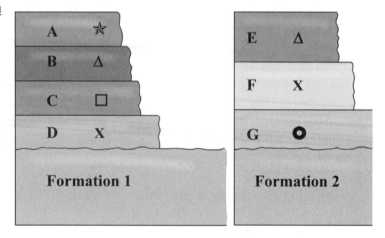

20. According to uniformitarians, which layer in Formation 1 represent the same time period as layer "E" in Formation 2?

21. According to uniformitarians, which layer in Formation 2 has no corresponding layer in Formation 1?

22. According to uniformitarians, which layer (out of all layers in both formations) has the oldest rock?

23. What is an intermediate link and why do paleontologists look for them?

24. Name two problems with the uniformitarian view of geology.

25. Name two problems with the catastrophist view of geology.

QUARTERLY TEST #3

1. Define the following terms:

a. Metabolism
b. Cell
c. Eukaryotic cell
d. Decomposers
e. Exoskeleton
f. Symbiosis
g. Herbivore
h. Basal metabolic rate

2. A scientist is examining a microscopic entity that targets cells, invades the cells, and forces the cells it has invaded to make duplicates of itself. It contains no DNA. Is it alive?

3. One strand of a DNA molecule has the following nucleotide base sequence:

guanine, adenine, cytosine, thymine

What will the sequence be on the same spot on the other strand?

4. The process of metabolism usually produces energy, carbon dioxide, and what else?

5. The process of photosynthesis requires energy from the sun, water, and what else?

6. You are examining an organelle that contains most of the cell's DNA. What organelle are you examining?

7. Blackbird chicks hatch blind and featherless. Are they precocial or altricial?

8. An organism is made of several eukaryotic cells and makes its own food. It also has roots, stems, and leaves. To which kingdom does it belong?

9. An organism is made of a single, prokaryotic cell. To what kingdom does it belong?

10. Why did people salt meat before refrigeration was widely available?

11. An organism is in kingdom Protista. It makes its own food and cannot move on its own. Is it a part of the protozoa or algae?

12. If a plant cell has no central vacuole, what will it be unable to maintain?

13. If a cell has distinct, membrane-bound organelles but not a cell wall, is it most likely a prokaryotic cell, an animal cell, or a plant cell?

14. What kind of muscle is found in the heart?

15. What is the name of the process that forms the top layers of your skin as well as your hair and nails?

16. Are the cells in the strands of your hair alive?

17. What do tendons do?

18. What two functions does sweat perform?

19. A person gets a mysterious disease that shuts down his sebaceous glands. Name one consequence of this situation.

20. In addition to fuel, what does combustion require?

21. If you were to weight the number of ounces of macronutrients you eat each day and compare that to the number of ounces of micronutrients you eat each day, which would be lower?

22. If you analyzed the specific proteins you eat and then looked for those exact same proteins in your cells, would you find them?

23. Two animals are roughly the same size and weight. They are kept in exactly the same conditions, and they are forced to have roughly the same amount of activity every day. Nevertheless, the first one eats significantly more than the second, but neither animal gains any weight. Which animal is ectothermic?

24. How many basic steps are there in the combustion process that takes place in your cells?

25. If you eat fewer calories than your BMR and activity level require, will you lose weight, gain weight, or stay at the same weight?

QUARTERLY TEST #4

1. Define the following terms:

a. Digestion
b. Vitamin
c. Arteries
d. Capillaries
e. Gland
f. Hormone
g. Autonomic nervous system
h. Sensory nervous system

Questions 2 through 7 are concerned with the following organs

stomach, esophagus, pancreas, large intestine, gall bladder, pharynx, small intestine, liver

2. Which organs are not part of the digestive tract?

3. Which organ produces sodium bicarbonate, which is a base?

4. In which organ does most of the absorption of nutrients take place?

5. Which organ produces bile?

6. Which organ does food pass through first, the pharynx or the esophagus?

7. Which organ concentrates bile?

8. Which side of the human heart has only deoxygenated blood in it?

9. If oxygenated blood is flowing through a vein, where has it been and where is it going?

10. Which structures in the lungs will have more capillaries: the bronchial tubes or the alveoli?

11. What chemical gives blood its red color?

12. When air is headed from your mouth to your lungs, does it pass through the larynx or trachea first?

13. While a woman is singing, she is changing her voice's pitch to match the melody. What is she doing to her vocal cords to change her voice's pitch?

14. Which system in the human body fights disease?

15. What is the main function of the urinary system?

16. You are watching microscopic video of two lymphocytes physically attacking a pathogen, trying to beat a hole in its membrane. Are you watching the work of B-cells or T-cells?

17. What piece of medical technology makes use of the body's ability to produce memory B-cells?

18. What nickname is often given to the pituitary gland in order to describe its function?

19. What is cerebrospinal fluid?

20. The two hemispheres of a person's brain are unable to communicate. What structure in the brain is not working?

21. Which side of the brain controls the left side of the PNS?

22. If a person's pupil opens as wide as possible, has he walked into a dark room or a well-lit room?

23. In #22, did the sympathetic division or the parasympathetic division of the person' autonomic nervous system make the change to the pupil?

24. A biologist is examining neurons from a person's brain under a microscope. He inadvertently spills some of that same person's blood onto the specimen. If she keeps watching the neurons, what will she see happening to them?

25. When you hear sound, which vibrates first: your ear drum or the fluid in the cochlea?

SOLUTIONS TO QUARTERLY TEST #1

1. a. (1 pt) <u>Spontaneous generation</u> – The idea that living organisms can be spontaneously formed from non-living substances

b. (1 pt) <u>Counter example</u> – An example that contradicts a conclusion

c. (1 pt) <u>Hypothesis</u> – An educated guess that attempts to explain an observation or answer a question

d. (1 pt) <u>Control (of an experiment)</u> – The variable or part of the experiment to which all others will be compared

e. (1 pt) <u>Double-blind experiments</u> – Experiments in which neither the participants nor the people analyzing the results know who is in the control group

f. (1 pt) <u>Simple machine</u> – A device that either multiplies or redirects a force

g. (1 pt) <u>Mechanical advantage</u> – The amount by which force or motion is magnified in a simple machine

2. (1 pt) Gregor Mendel's greatest contribution to science was <u>his work on how traits are passed on during reproduction</u>. The student could also say "genetics."

3. (1 pt) Max Planck first made the assumption <u>that energy comes in small packets called "quanta."</u>

4. (1 pt) Copernicus championed the <u>heliocentric</u> system, which is the best description of how the planets and sun are arranged.

5. (1 pt) We call them <u>alchemists</u>.

6. (1 pt) Those are the accomplishments of <u>Newton</u>.

7. (1 pt) <u>No</u>, you should not believe it because Dr. Collins says so. Even the greatest scientists can be wrong. You must believe or reject it because of what the data say.

8. (1 pt) <u>It is a theory</u>. It was a hypothesis before the tests were done, and it will take generations of data to turn it into a law.

9. (1 pt) <u>Neither</u> will hit the bottom of the chamber first, as all things fall at the same rate in the absence of air.

10. (1 pt) <u>No</u>, science cannot prove anything.

11. (1 pt) <u>The scientist has found data that are not consistent with the hypothesis</u>. He might have done the experiments himself, or he might have found some other scientist's data. However, if he is faced with the prospect of discarding or modifying his theory, it must not be consistent with the data.

12. (1 pt) Lowell is best known for his theory that <u>there are canals on Mars</u>.

13. (1 pt) <u>Yes</u>, as long as the scientific method is followed, science can be used to study *anything*!

14. (1 pt) <u>Group A</u> is the control, because they are not getting the real drug.

15. (1 pt) This should be a <u>double-blind</u> experiment, since the people could be affected by thinking they are getting a cold-prevention drug, and since the data being collected are subjective.

16. (1 pt) If fewer people got sick in group B, that gives you evidence that the drug is <u>effective</u>, at least to some extent.

17. (1 pt) The weatherman's office is represented by the temperature that is 0 miles from the office. It is the <u>hottest of all the temperatures</u>.

18. (1 pt) At 6 miles from the office, the average temperature was <u>58</u> °F.

19. (1 pt) You should be <u>somewhere between 3 and 5 miles</u> from his office. The student can list the range or just give a single value within that range.

20. (2 pts – one for the correct equation, one for the correct answer) For levers, the mechanical advantage equation is:

Mechanical advantage = (distance from fulcrum to effort) ÷ (distance from fulcrum to resistance)

Mechanical advantage = 100 ÷ 10 = <u>10</u>

21. (1 pt) In first-class levers, the mechanical advantage <u>magnifies the force</u>.

22. (1 pt) In a block and tackle, the mechanical advantage is the number of pulleys, which is <u>4</u>.

23. (1 pt) You will have to pull 4 times as much rope, so you will have to pull <u>40 feet</u> of rope.

24. (1 pt) Use a <u>fatter</u> one, as it is the circumference that matters in calculating the mechanical advantage.

25. (1 pt) They were known for <u>technology</u>. They didn't do science because they didn't try to explain the world around them. They didn't do applied science, because they didn't conform to the scientific method. They did have a lot of things (like great medical practices) that made life better, though, and that is technology.

Total points: 32

SOLUTIONS TO QUARTERLY TEST #2

1. a. (1 pt) <u>Aristotle's dictum</u> – The benefit of the doubt is to be given to the document itself, not assigned by the critic to himself.

b. (1 pt) <u>Dendrochronology</u> – The process of counting tree rings to determine the age of a tree

c. (1 pt) <u>Minerals</u> – Inorganic crystalline substances found naturally in the earth

d. (1 pt) <u>Unconformity</u> – A surface of erosion that separates one layer of rock from another

e. (1 pt) <u>Petrifaction</u> – The conversion of organic material into rock

f. (1 pt) <u>Resin</u> – A thick, slowly flowing liquid produced by plants that can harden into a solid

g. (1 pt) <u>Index fossils</u> – Fossils that are assumed to represent a certain period in earth's past

h. (1 pt) <u>Geological column</u> – A theoretical picture in which layers of rock from around the world are meshed together into a single, unbroken record of earth's past

2. (1 pt) The <u>external</u> test compares manuscripts to other known historical and archaeological facts.

3. (1 pt) The <u>first</u> book is more likely to be true to the original from which it was copied, because it has the shorter time span between copy and original and also the larger number of supporting documents.

4. (1 pt) The coin has a <u>known</u> age.

5. (1 pt) He decides it is <u>younger</u> than the coin. Since it is in a layer of soil above the coin, he assumes the coin had to have been preserved first.

6. (1 pt) The archaeologist is using the <u>Principle of Superposition</u>.

7. (1 pt) <u>There are many seemingly unrelated cultures that all have a worldwide flood tale</u>. If the flood did not really occur, you have to assume that they all made up the tale independently, because many of the cultures had no contact with one another until well after the tales were written down.

8. (1 pt) <u>Catastrophism</u> is more open-minded, as it does not lock you into any presupposed age for the earth.

9. (1 pt) <u>Igneous rock</u> is the result of magma that cools and solidifies.

10. (1 pt) This is <u>chemical weathering</u>. The limestone forms a gas. That changes the composition of what's left.

11. (1 pt) An unconformity between two parallel strata of sedimentary rock is a <u>disconformity</u>.

12. (1 pt) This is a <u>paraconformity,</u> and it is a desperate attempt to keep the uniformitarian view consistent with the index fossils found in the rock.

13. (1 pt) You would expect less erosion in the <u>forest,</u> because trees and plants tend to hold soil together.

14. (1 pt) Fossils are generally found in <u>sedimentary</u> rock.

15. (1 pt) <u>No,</u> you cannot. You can't be 100% sure of anything from science, and we know that there are several cases in which scientists thought a plant or animal was extinct and then were shocked to find a living specimen.

16. (1 pt) Most likely, the animal will be <u>petrified.</u>

17. (1 pt) You have found the <u>carbonized remains</u> of a plant.

18. (1 pt) According to uniformitarians, each layer of rock represents a time period in earth's history. Thus, <u>what organisms existed during that time period</u> determined what fossils are in that layer of rock.

19. (1 pt) According to catastrophists, <u>the stage of the Flood and what creatures were trapped at that stage</u> determined what fossils are in that layer of rock.

20. (1 pt) Since layer "<u>B</u>" has the same index fossils as layer "E," it represents the same time period.

21. (1 pt) Since no layer in Formation 1 has the same index fossils as layer "<u>G</u>" in Formation 2, Formation 1 has no layer corresponding to that time period.

22. (1 pt) Layer "<u>G</u>" contains the oldest rocks. Layer "D" is at the bottom of Formation 1, indicating it is the oldest layer in that formation. However, layer "F" has the same index fossils, meaning it represents the same time period. Since "G" is under "F," then, it must be the oldest.

23. (2 pts – one for what a fossils is and one for why paleontologists look for them) <u>An intermediate link is a fossils that shows characteristics indicative of two different kinds of animals.</u> For example, a fossil of a fish-like creature with legs would be an intermediate link between fish and amphibians. <u>Paleontologists look for them because they are expected if the Theory of Evolution is true.</u> The fact that no unambiguous examples can be found indicate that the theory is not valid.

24. (2 pts – the student need list only two of the following) The text discusses the following problems:

a. <u>There are too many fossils in the fossil record.</u>

b. <u>Fossils such as the *Tyrannosaurus rex* bone that contains soft tissue are hard to understand in the uniformitarian framework.</u>

c. <u>Fossil graveyards with fossils from many different climates are hard to understand in the uniformitarian view.</u>

d. Index fossils are called into question by the many creatures we once thought were extinct but we now know are not.

e. Uniformitarians must assume the existence of paraconformities.

f. Uniformitarians must believe that evolution occurred, and there is no evidence for evolution. In fact, the fossil record provides evidence that each plant and animal was created by God.

25. (2 pts – the student need list only two of the following.) The text discusses the following problems:

a. Catastrophists have offered no good explanation for the existence of unconformities between rock layers laid down by the Flood.

b. Catastrophists cannot explain certain fossil structures that look like they were formed under "normal" living conditions which would not exist during the Flood.

c. Catastrophists have not yet explained the enormous chalk deposits we find in terms of the Flood.

Total points: 35

SOLUTIONS TO QUARTERLY TEST #3

1. a. (1 pt) Metabolism – The sum total of all processes in an organism that convert energy and matter from outside sources and use that energy and matter to sustain the organism's life functions

b. (1 pt) Cell – The smallest unit of life in creation

c. (1 pt) Eukaryotic cell – A cell with distinct, membrane-bounded organelles

d. (1 pt) Decomposers – Organisms that break down the dead remains of other organisms

e. (1 pt) Exoskeleton – A body covering, typically made of chitin, that provides support and protection

f. (1 pt) Symbiosis – A close relationship between two or more species where at least one benefits

g. (1 pt) Herbivore – A consumer that eats producers exclusively

h. (1 pt) Basal metabolic rate – The minimum amount of energy required by the body in a day

2. (1 pt) It is not alive. It must have DNA to be alive. The description given is actually consistent with certain kinds of viruses.

3. (2 pts, ½ for each) Since we know thymine and adenine link together and guanine and cytosine link together, we know the other strand must have:

cytosine, thymine, guanine, adenine

4. (1 pt) water

5. (1 pt) carbon dioxide

6. (1 pt) The majority of DNA is stored in the cell's nucleus.

7. (1 pt) They are altricial, because they cannot see or regulate their body temperature on their own.

8. (1 pt) Since it is made up of several eukaryotic cells, it is not in kingdom Monera. Since it makes its own food, it is probably a plant. However, you have to be a bit concerned, because it might be an alga and thus belong to kingdom Protista. However, algae do not have specialized structures, so this is not an alga. It therefore must be in kingdom Plantae.

9. (1 pt) If it is a single, prokaryotic cell, it must be in kingdom Monera.

10. (1 pt) The presence of salt reduces the growth and reproduction of bacteria. Thus, salt protects meat from contamination by bacteria.

11. (1 pt) It is a part of the algae.

12. (1 pt) It will be unable to maintain turgor pressure.

13. (1 pt) It is most likely an <u>animal</u> cell. Prokaryotic cells don't have distinct, membrane-bound organelles, and plant cells have cell walls.

14. (1 pt) <u>Cardiac</u> muscle is in the heart.

15. (1 pt) <u>Keratinization</u> is the process that hardens living cells. It is used to make the outer layer of the epidermis, as well as hair and nails.

16. (1 pt) <u>No</u>, keratinization kills cells.

17. (1 pt) Tendons <u>attach skeletal muscles to the endoskeleton.</u>

18. (2 pts – one for each function) Sweat <u>cools the body down and also helps feed the beneficial bacteria and fungi that live on your skin.</u>

19. (1 pt – the student need list only one). Since the sebaceous glands produce oil that softens the skin and hair, you would have <u>dry skin</u>. The student could say things like "chapped skin" as well. You could also have <u>wiry, brittle hair</u>. In addition, it helps repel certain pathogens, so you could <u>experience more infections</u>.

20. (1 pt) In addition to fuel, combustion requires <u>oxygen</u>.

21. (1 pt) You eat a *lot* fewer <u>micronutrients</u> each day.

22. (1 pt) You <u>would not</u> find any of those proteins in your cells. The proteins you eat are broken down into amino acids so your cells can build the proteins *they* need.

23. (1 pt) The <u>second</u> animal is ectothermic. The first animal needs to eat more because endothermic animals have a much higher BMR.

24. (1 pt) There are <u>three</u> basic steps – glycolysis, the Krebs cycle, and the electron transport system.

25. (1 pt) You will <u>lose</u> weight, because you are using more food than you are eating.

Total Points: 34

SOLUTIONS TO QUARTERLY TEST #4

1. a. (1 pt) <u>Digestion</u> – The process by which an organism breaks down its food into small units that can be absorbed by the body

b. (1 pt) <u>Vitamin</u> – A chemical substance the body needs in small amounts to stay healthy

c. (1 pt) <u>Arteries</u> – Blood vessels that carry blood away from the heart

d. (1 pt) <u>Capillaries</u> – Tiny, thin-walled blood vessels that allow the exchange of gases and nutrients between the blood and cells and are located between arteries and veins

e. (1 pt) <u>Gland</u> – A group of cells that prepare and release a chemical for use by the body

f. (1 pt) <u>Hormone</u> – A chemical messenger released into the bloodstream that sends signals to distant cells, causing them to change their behavior in specific ways

g. (1 pt) <u>Autonomic nervous system</u> – The system of nerves that carries instructions from the CNS to the body's smooth muscles, cardiac muscle, and glands

h. (1 pt) <u>Sensory nervous system</u> – The system of nerves that carries information from the body's receptors to the CNS

2. (1 pt – one-third for each) The <u>pancreas</u>, <u>gall bladder</u>, and <u>liver</u> are not members of the digestive tract, because food does not pass through them.

3. (1 pt) The <u>pancreas</u> produces sodium bicarbonate.

4. (1 pt) Most nutrients are absorbed in the <u>small intestine</u>.

5. (1 pt) The <u>liver</u> produces bile.

6. (1 pt) The food must travel through the <u>pharynx</u> to get to the esophagus.

7. (1 pt) The <u>gall bladder</u> concentrates bile.

8. (1 pt) The <u>right</u> side of the heart receives deoxygenated blood and sends it to the lungs.

9. (1 pt – ½ for each) If it is a vein, the blood is <u>going to the heart</u>. The only time oxygenated blood flows towards the heart is when it <u>has come from the lungs</u>.

10. (1 pt) Capillaries will be where gas exchange takes place, and that happens at the <u>alveoli</u>.

11. (1 pt) <u>Hemoglobin</u> gives blood its red color. Give the student ½ a point if he says red blood cells.

12. (1 pt) It passes through the <u>larynx</u> first.

13. (1 pt) The woman changes the <u>tightness</u> of her vocal cords.

14. (1 pt) The <u>lymphatic system</u> fights disease.

15. (1 pt) The urinary system <u>regulates water balance and chemical levels in the blood</u>.

16. (1 pt) <u>T-cells</u> attack pathogens directly. B-cells make antibodies.

17. (1 pt) A <u>vaccine</u> uses this ability to make a person immune to a disease before the person is exposed to it.

18. (1 pt) It is often called the <u>master endocrine gland</u>.

19. (1 pt) It is <u>the fluid in which the brain floats</u>. It provides nutrition and support to the brain.

20. (1 pt) This person's <u>corpus callosum</u> is not working.

21. (1 pt) The <u>right</u> side of the brain controls the left side of the PNS.

22. (1 pt) He has walked into a <u>dark</u> room, since the pupil is trying to get as much light as possible into the eye.

23. (1 pt) The <u>sympathetic division</u> of the autonomic nervous system caused that change.

24. (1 pt) The biologist would see the neurons <u>die</u>, because blood is toxic to them. The student doesn't have to say "die." He could say, "get sick," or anything else that indicates exposure to a toxic substance.

25. (1 pt) The <u>ear drum</u> vibrates, which then causes the ossicles to vibrate, which then cause the fluid in the cochlea to vibrate.

Total points: 32